Cutting and Packing in Production and Distribution

Contributions to Management Science

Ulrich A. W. Tetzlaff
Optimal Design of Flexible Manufacturing Systems
1990. 190 pages. Softcover DM 69,-
ISBN 3-7908-0516-5

Fred von Gunten
Competition in the Swiss Plastics Manufacturing Industry
1991. 408 pages. Softcover DM 120,-
ISBN 3-7908-0541-6

Harald Dyckhoff · Ute Finke

Cutting and Packing in Production and Distribution

A Typology and Bibliography

With Support from Ute Vieth

With 14 Figures

Springer-Verlag Berlin Heidelberg GmbH

Series Editors
Werner A. Müller
Peter Schuster

Authors
Univ.-Professor Dr. Harald Dyckhoff
Dipl.-Ök. Ute Finke
Lehrstuhl für Industriebetriebslehre
RWTH Aachen
Templergraben 64
D-5100 Aachen, FRG

ISBN 978-3-642-63487-1

CIP-Titelaufnahme der Deutschen Bibliothek

Dyckhoff, Harald:
Cutting and packing in production and distribution : typology and bibliography / Harald Dyckhoff; Ute Finke. With support from Ute Vieth. - Heidelberg : Physica-Verl., 1992
(Contributions to management sciece)
ISBN 978-3-642-63487-1 ISBN 978-3-642-58165-6 (eBook)
DOI 10.1007/978-3-642-58165-6
NE: Finke, Ute:

Originally published by Physica-Verlag Heidelberg in 1992
Softcover reprint of the hardcover 1st edition 1992

7120/7130-543210

Preface

Cutting and packing problems such as the cutting of sheet metal and the loading of containers or, in a more abstract sense, capital budgeting or assembly line balancing have been treated in scientific literature of various disciplines for about fifty years. Since the pioneer work of Kantorovich in 1939, which first appeared in the West in 1960, there has been a steadily growing number of contributions of increasing importance in this research area, particularly in the past decade. As of today more than 700 contributions exist even when applying a strict categorisation. Since comprehensive monographies and compiled studies are still lacking, it is very time consuming and thus expensive to search for a suitable solution procedure for concrete problems in the available literature. Thus, an apparently simpler way is often chosen, which is to develop ones own approach. For this reason there is not only the danger of unnecessary effort and scientific repetition, but it is reality.

With the goal of improved research coordination an interdisciplinary Special Interest Group on Cutting and Packing (SICUP) was founded in 1988, which meets every two years (1988 Paris, 1990 Athens, 1992 San Francisco) and issues a semi-annual newsletter (SICUP-Bulletin) with up-to-date information.

This book is intended to assist in the coordination of research work in this area. Prompted by frequent inquiries from interested researchers and practitioners for references to literature or specialists dealing with specific problems, and based on the experience that relevant sources are too often overlooked in publications, we decided to analyze existing knowledge about relevant cutting and packing problems in a systematic form, hopefully avoiding unnecessary work in the future. The result of these efforts is the presented typology for cutting and packing problems, which is based on their logical structure.

The description offers both theoretical insight and practical possibilities for application. On a more theoretical level, a systematic catalogue of properties and a respective hierarchical catalogue of types are made available as instruments, with which the numerous problems of cutting and packing can be characterized and systematized in respect to their essential similarities and differences. Moreover, they provide suggestions for a consistent terminology. On a more practical level, the detailed description of over 300 problems published in the scientific literature provides an appropriate and easy access to existing knowledge. Additionally, an appendix gives a rough systematization of approximately 400 further sources. All the data is contained in a literary data base and can be acquired as an ASCII-file from the authors for a nominal charge.

As a whole, we think that this book is suitable for serving as a starting point in the topic of cutting and packing problems. Interested specialists may receive new stimulus in the solving of these problems and, finally, be able to use the book for further reference.

The investigation presented is the result of a two-year research project, concluded in June 1991, which was financially supported by the Deutsche Forschungsgemeinschaft (DFG). We are very grateful for their contribution.

We would also like to express a special appreciation to all who have taken part in the completion of this book. In the evaluation of literary sources and editorial tasks, the support of Dipl.-Kff. Ute Vieth has proved to be so valuable that we have mentioned her assistance in the bibliographical notes. Special thanks also belongs to Prof. Dr. Gerhard Wäscher for his interest and constructive criticism in this project and for his numerous suggestions. We are also grateful to Dr. Eberhard Bischoff and Dr. Hermann-Josef Kruse for their helpful advice and productive remarks at various stages of the manuscript draft. We naturally assume full responsibility for any possible (hopefully few) mistakes or inconsistencies in the final draft. Among the industrious assistants who have participated in completing this book, we would like to particularly thank Dipl.-Kfm. Dirk Kall for his help in the task of searching for sources, registering them in the data base and for his conscientous completion of the tables, as well as our student Andrea Dietel for her patient typing and retyping of the final manuscript. All of them deserve our best thanks.

Aachen, April 1992
Harald Dyckhoff and Ute Finke

Table of Contents

1. Introduction

1.1. Purpose of the Investigation

Within such disciplines as Management Science, Information and Computer Sciences, Engineering, Mathematics and Operations Research, problems of cutting and packing (C&P) of concrete or abstract objects appear under various specifications (cutting problems, knapsack problems, container and vehicle loading problems, pallet loading, bin packing, assembly line balancing, capital budgeting, changing coins, etc.), although they all have essentially the same logical structure. When cutting, a large object must be divided into smaller pieces; when packing, small items must be combined to large objects. In this way, packing problems, such as loading containers, may also be looked at as those of cutting: The empty space of a container must be divided (or 'cut') into smaller parts, determined by packages which have to be loaded. In the same way the problem of cutting solid material can be identified with packing an empty space. An example is in the textile industry, where paper templates of the pieces of clothing to be cut are used to find an optimal material utilization. With the templates as substitutes for the figure of the pieces of clothing, a packing problem appears: the templates must be packed on the textile strips. Whether a problem is seen as one of cutting or packing thus is simply a question of the point of view. Either one considers the geometric bodies of objects or the space occupied by them.

Because of the strong link between cutting and packing problems based on the duality of material and space, it seems to be obvious to examine both within a general framework. Statements made concerning cutting problems can also be significant for problems in packing, and vice versa.[1]

This does not mean, however, that one can concentrate on one of the aspects while neglecting the other. Each has its own characteristics which must be considered. For instance, as to the applicability of the so-called "guillotine cut", where an uninterrupted cut is made from one end of the object to the other, there exists a crucial difference between cutting and

1) For this conception see also Eilon (1960), Brown (1971), Golden (1976), Müller-Merbach (1981).

packing problems. When cutting objects, this technique is often technologically unavoidable or at least desirable, whereas in packing it may cause instability during transportation.

Despite a constant growth of scientific discussion on C&P problems - more than 700 scientific studies have been published - remarkably few have attempted to examine the similarities systematically, providing conclusions which could then be applied to specific situations. In particular, numerous solution concepts or recommendations developed so far are insufficiently accessible for practical use. Existing surveys[1] have generally concentrated on specific mathematical or technical aspects of C&P problems. What is necessary, however, is a description and categorization of C&P problems as a whole, pointing out their essential similarities and differences. This enables a systematic comparison of all the various kinds of problems and suggested solution approaches and contributes to better use of the published knowledge on C&P problems.

Necessary requirements for providing a comprehensive picture in the field of C&P problems are:

(1) Creation of a consistent terminology for C&P problems

In order for the various C&P problems to be compared, it is necessary to have a consistent terminology. Up to this point, this has not been the case. As an example, in Computer Science and Mathematics the concept of 'bin-packing', which was originally developed for memory allocation, is presently used to include such notions as vehicle loading and cutting of glass, textiles, etc. On the other hand, in Operations Research, Management Science and Industrial Engineering both the terms 'loading problem' and 'cutting stock problem' comprise bin packing problems. A second example is in applying the term 'cutting stock problem'. In the practical literature this problem is described as a task of cutting a <u>lot</u> of large objects into smaller items, whereas in the methodical literature a 'general cutting stock problem' occurs simply when <u>one</u> large object is to be divided. Within the terminology proposed in this book the last case will be defined as a (multi-dimensional) knapsack problem.

(2) Development of an appropriate characterization scheme

In order to adapt C&P solution methods for practical usage, it is necessary to develop criteria which crystallize the logical structure of these problems despite their apparent differences. Thus, a quick access to existing knowledge is ensured. This will help to avoid

1) E.g. Brown (1971), Golden (1976), Hinxman (1980), Garey-Johnson (1981), Rayward-Smith/Shing (1983), Coffman et al. (1984), Dowsland (1985b), Terno et al. (1987). See also Table A.1, p. 164.

unnecessary repetition in scientific work, which has unfortunately been a reality in this area of investigation.

(3) Systematic survey of literature

In concrete situations and when developing standard software existing solution approaches can be an important help. However, to locate the appropriate procedure in the existing literature - if possible at all - requires much time and is therefore such an important cost factor that people often create their own procedure.[1] If, however, C&P problems researched so far are characterized using a detailed catalogue of properties and are accessible in form of a systematic literary survey, the probability of locating similar types of problems with suitable solution and handling procedures will increase.

These three points are the main goals of this book. The purpose is to create a basis of knowledge which provides suggestions for the selection, development and utilization of appropriate solution approaches and perhaps to provide a basis for an expert system which can be developed later.

1.2. Methodology Used

The starting point of our investigation was the basic logical structure of C&P problems. This structure allowed for a first, rough arrangement of concrete phenomena and elementary problem types, and was applied to specify the area of investigation. Additionally, it served as a foundation for determining characteristic properties in problems of cutting and packing. Two categories appear,

- characteristics based on the logical structure and
- characteristics based on reality.

Based on findings and experience of earlier research[2] a catalogue of attributes[3] was developed, later on tested according to its practical relevance, the results modified and then reduced to a few essential ones. The catalogue thus consists of 34 attributes with a maximum of four different properties each.

1) The existing literature reviews compiled by Madsen (1980), Dyckhoff/Wäscher (1988), and Sweeney (1989) are only of little help for this purpose due to lack of structurization.

2) Cf. Dyckhoff (1987) as well as Dyckhoff et al. (1985), (1988a) and (1988b).

3) Parts of the catalogue have already been published (Dyckhoff, 1990 and 1991b). This investigation will refine, modify and expand on the catalogue.

Parallel to this work, an extensive bibliography of available literature on C&P problems was gathered and systematically stored in a data base.[1]

The sources were analysed according to our catalogue of attributes and the resulting properties were also stored in the data base.[2] Thus, a 'fund' of possible characterizations of C&P problems in the current literature has become available.

Within the scope of this book we were interested in defining types of C&P problems which serve as a decision aid for the selection, development and utilization of solution procedures. The definition of the various problem types, therefore, is based on similarities in terms of solution approaches. Properties are combined in clusters and assigned to types which can be regarded as representing the essential characteristics of the solution to a group of C&P problems. It is important in this context that problems belonging to the same type should have similar properties in terms of the method of solution and that different types should have clear distinctions in this respect. Therefore, the goal of type-defining was not to use all developed attributes, but to isolate those which are significant to the solutions of C&P problems. Their combination (so-called **type-defining** characteristics) allowed for the definition of types, which on the one hand determine a variety of important characteristics and on the other hand still represent a great number of the actual phenomena of cutting and packing described in the existing literature. In this way it was possible to organize the abundant sources of C&P problems and to produce an overall view thereof.

The actual definition of types was organized hierarchically under the aspects of abstraction and aggregation. On the basis of the properties found in the C&P literature, and through "intuition and construction"[3] (a constant juggling between empirical observation and mental classification on the one hand and considerations of plausibility factors and testing of validity on the other), a hierarchical catalogue of C&P problem types was developed. On the first level of the hierarchy so-called **general types** are differentiated. These are phenomena which were constructed using the smallest possible quantity of type-defining characteristics. When possible or necessary, these types were categorized into **special types**, in partially various degrees of abstraction, by combining them step by step with additional type-defining properties. This process resulted in a systematic development of type-chains, through which each type fully describes a C&P problem with various degrees of abstraction. The differentiation process stops if there are no more type-defining properties. The existing C&P prob-

1) Cf. Dyckhoff/Wäscher (1988). The bibliography by Sweeney proved helpful by completing our own data base with additional 50 sources.

2) In order to limit the extent of the investigation it was necessary to exclude some sources, such as (masters') theses, articles in some periodicals for practitioners and many purely mathematical studies, particularly those published on bin-packing and knapsack problems; see also Section 3.3.

3) Cf. Große-Oetringhaus (1974) p. 34. An attempt was also made to apply cluster-analytical approaches to encourage suggestions for tracing reality-based types.

lems described in literature were assigned to the final links of these type-chains. These C&P problems were examined in respect to those characteristics which were not used for definition of types (so-called **type-describing** characteristics).

1.3. Structure of the Book

In **Chapter 2** C&P problems will be examined in an abstract manner as geometric-combinatoric problems, whereby the fundamental logical structure is explained. This structure is then used in a first, rough systematization of actual C&P problems, thus reducing the area of investigation.

Chapter 3 takes a closer look at the literature on C&P problems. First, the formulation of C&P problems into decision models will be discussed in order to show to what extent this sort of model can be understood as a picture of phenomena of cutting and packing. Then a systematization and delimitation of sources of C&P problems will follow, according to contextual and bibliographical criteria.

The purpose of **Chapter 4** is to present the characteristics used to analyse C&P problems. In this chapter, the properties will be individually described with explanations for their selection.

These properties lead to the definition of abstract types in **Chapter 5**, which group C&P problems described in literature according to aspects relevant in decision-making. These 'real problems' will be characterized according to type-defining properties and then analyzed in a more detailed fashion according to type-describing properties **from Chapters 6 to 9**.

Chapter 10 presents the summarized assessment of the results of this investigation with regard to 'theoretical insight' and 'practical application'.

Following the report is an extensive appendix which includes:

> **Appendix I** : A supplementary bibliography of C&P sources which are not used in the book. Part A lists contributions which give an overview of problems and/or techniques for specific types of C&P problems. In Part B there are sources which examine abstract or related problem areas, introduce bin-packing or knapsack problems on a mathematical basis, or describe software solutions. Furthermore, 'non-scientific' sources are listed, published in some magazines for practitioners as well as discussion and working papers having appeared in 1986 or later. Part C lists all kinds of sources which have appeared or have become available to the authors after the editorial deadline (beginning of 1991).

Appendix II : Brief descriptions of characteristic properties presented thoroughly in Chapter 4.

Appendix III : Short illustration of the data base system "LARS" (Leistungsstarkes Archivierungs- und Recherchesystem) and description of the structure of the 'C&P' data base.[1]

1) Access to the data base can be achieved by contacting the authors and requires the availability of the LARS programme (cf. Appendix A. III).

2. Cutting and Packing Problems as Geometric-Combinatoric Problems

In order to make C&P problems comparable in respect of their solutions, it is useful to steer away as much as possible from their practical setting and to concentrate on the common, abstract foundations of the problems. This provides a common basis and structural foundation for researching various concrete phenomena.

2.1. Basic Logical Structure

The basic logical structure of C&P problems becomes obvious through the consideration of some simple examples:[1]

Figure 2.1 illustrates a problem of cutting tubes for radiators.[2] The segments required (lengths 5 to 46) are to be cut from the available tubes of length 98. For this task a way is sought which can be applied for cutting the tubes in the most efficient way. Under consideration of certain objectives and restrictions, the lengths of the ordered tubes are combined to create a pattern, which is then assigned as a rule for cutting the large tubes on stock. Since the length of the large tubes is rarely equal to the sum of a combination of lengths of the smaller tubes, residual pieces will result, so-called trim loss.

In a container loading problem[3], illustrated in Figure 2.2, there are various-sized packages which must be packed into several stock containers of one size. Apart from certain individually given objectives and restrictions, the fundamental packing problem concerns combining the small items into a pattern, which then serves as a prescription for the assignment of the large containers on stock. Since the measurement of the combined packages generally does not equal that of the large containers, one can assume that unused space is left. This, too, can be defined as trim loss.

---

1) Cf. Dyckhoff (1990) pp. 145 ff.

2) Cf. Heicken/König (1980).

3) Cf. Liu/Chen (1981).

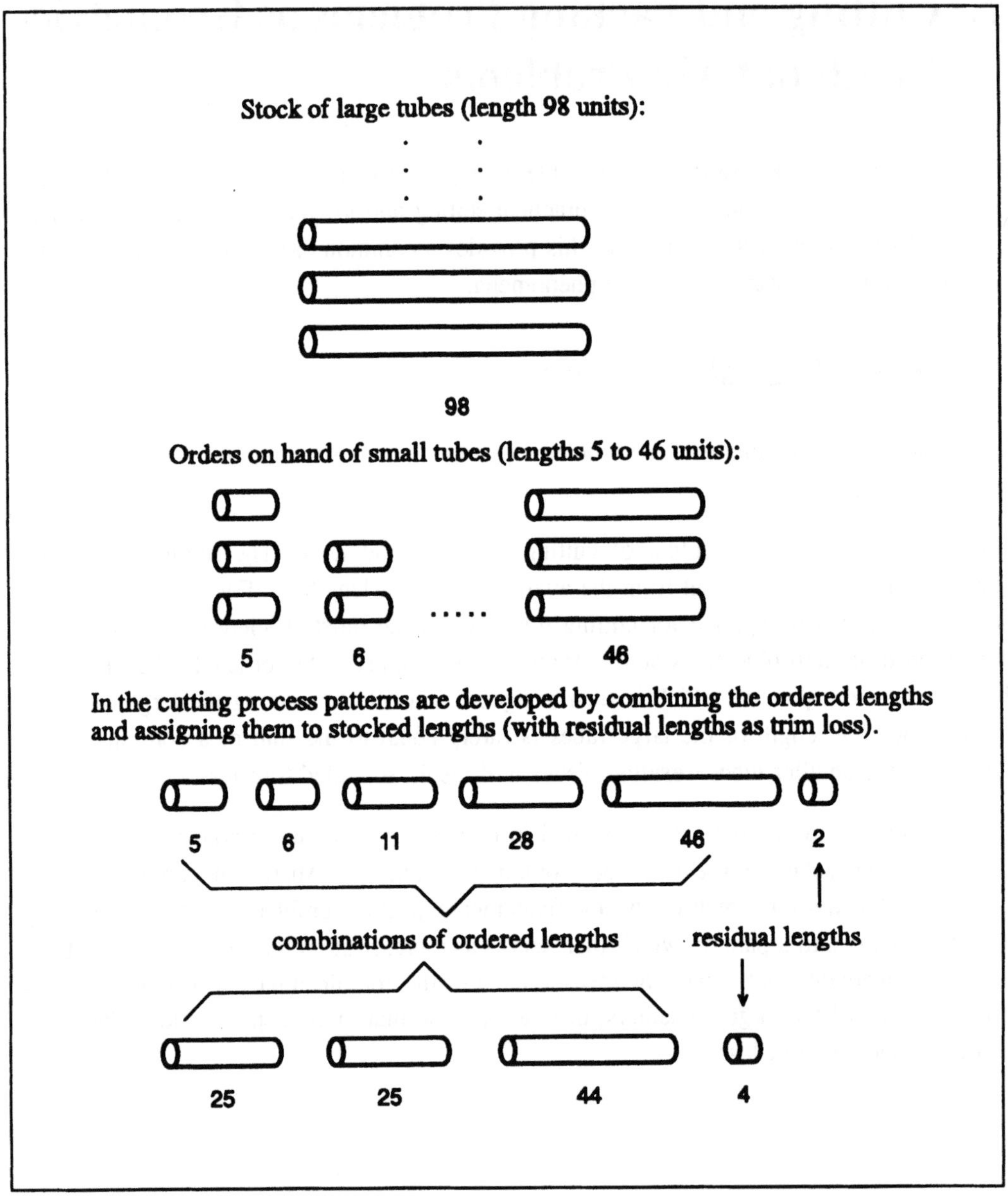

Figure 2.1: Tube cutting

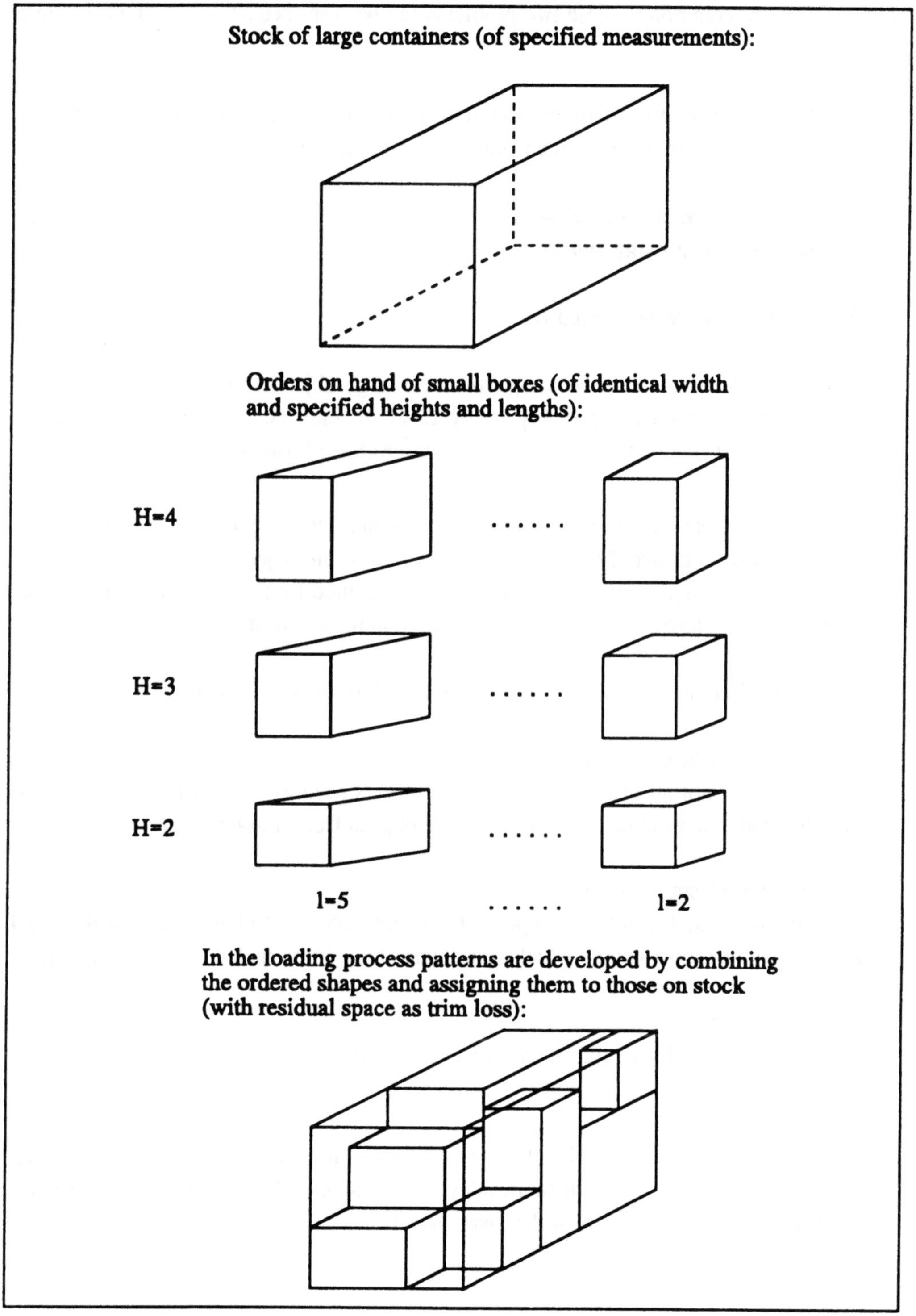

Figure 2.2: Container loading

Thus, the features common to these two problems can be described as following (see Figure 2.3):[1]

(1) There are two groups of objects represented as geometrical bodies of specific measurements in a finite-dimensional space of real numbers:

- a given stock of **large objects** and
- a given list of **small items.**

(2) Two tasks are to be carried out:

(a) Small items are to be **assigned** to the large objects. Each large object is assigned a given set, possibly empty, of small items, and each item is assigned to at most one of the large objects (**combinatoric base condition**).

(b) Within each large object, one or more small items are to be **arranged** in such a way as to avoid overlapping and to fit into the object's geometrical boundaries. The space of the object which is not filled by the orders on hand is normally referred to as **trim loss (geometric base condition**).

There are several components to this **geometric-combinatoric problem**, such as:[2]

(1) Selection of large objects
In many cases, not all of the available large objects need to be used for a given order. The first component is the selection of large objects from a given stock.

(2) Selection of small items
In C&P problems, it is not always possible or necessary to fulfill all orders for the small items. Thus one must decide which small items have to be included in the process.

(3) Assignment
One must decide which small items are to be assigned to a large object.

(4) Layout
Within every selected large object there is naturally the question concerning the actual arrangement of small items. In short, a layout, or so-called (cutting or packing) pattern, must be determined for each of the large objects.

1) Cf. Dyckhoff (1987) pp. 104 and (1990) pp. 146 ff.

2) Cf. Dyckhoff (1987) pp. 105 ff., Wäscher (1989b) p. 41.

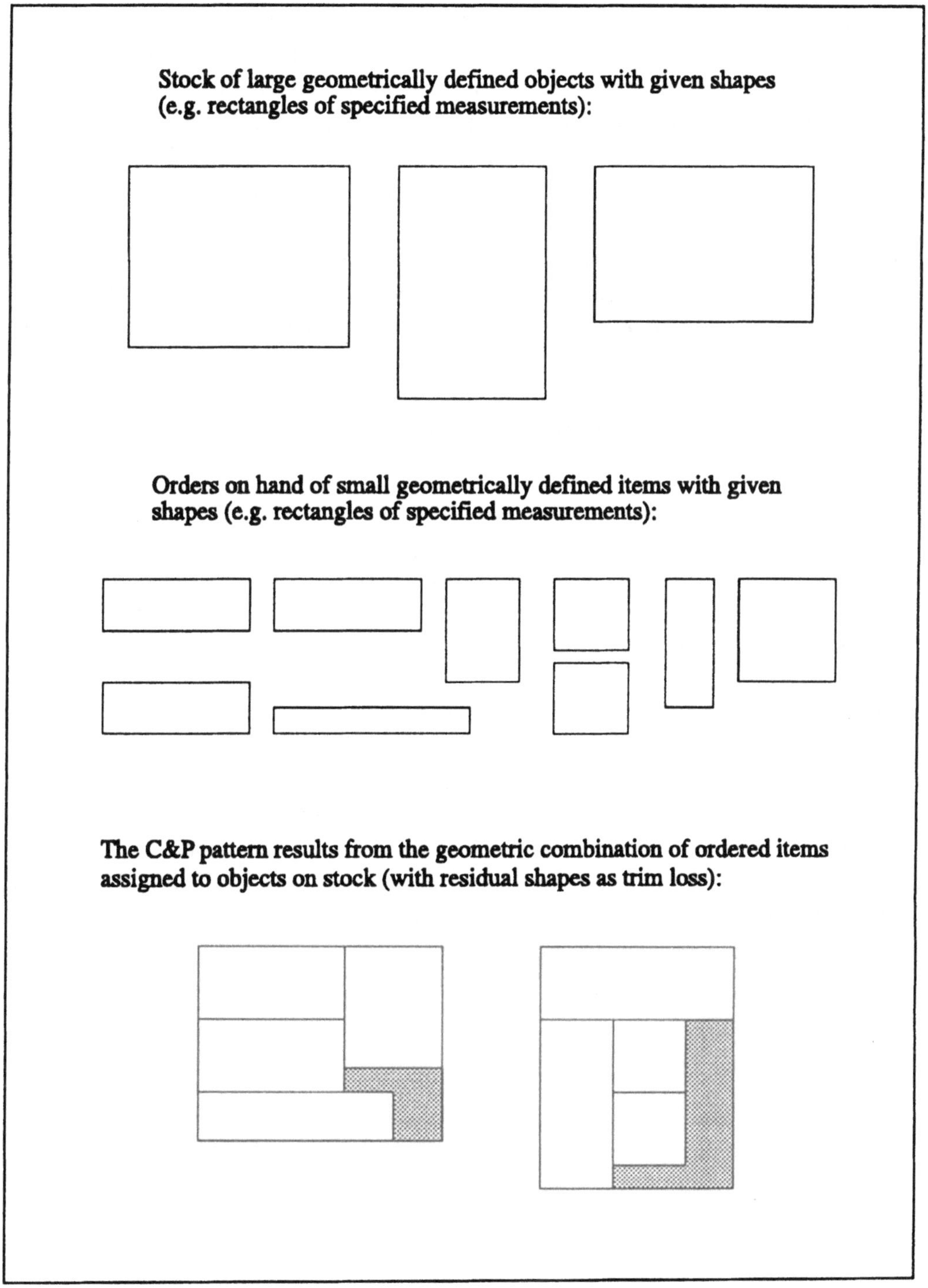

Figure 2.3: Basic logical structure of C&P problems

In a concrete C&P problem it is not always necessary that each of these points comes into question. For example, if small boxes of identical measurements should be packed onto a pallet, the selection of large objects (i.e. pallets) is trivial. On the other hand it is quite common in C&P problems that additional decisions need to be made - for example in the textile industry where the fabric is layered for cutting[1] - but since these points are specific to the practical setting they are not included in the basic structure of these problems.[2]

It shall be emphasized that the subdivision of C&P problems into components is based on imagination and only serves a better understanding of the nature of the problem. In finding a solution to a C&P problem it is necessary to consider all relevant aspects simultaneously.

2.2. Phenomena of Cutting and Packing

Since the definition of C&P problems as geometric-combinatoric problems is orientated on logical structure rather than on actual phenomena, it is possible to attribute apparently heterogenous problems to common backgrounds and to recognize general similarities. Thus, in a broader sense not only material processes are associated with C&P problems, but also problems concerning abstract objects. An example is in assembly line balancing[3] (see Figure 2.4), where tasks of specified durations are to be completed in a certain sequence and assigned to ("packed into") the fewest possible work stations. Thus, the individual tasks represent the small items and the stations the large objects. Important is that the objects and items can be distinguished by certain characteristics, which are relevant to the assignment and have comparable scales of measurement. With problems involving materials, the scale is usually the shape of the objects and items, whereas with abstract problems, the dimensions relevant to assignment can be based on time (in assembly line balancing the duration of the task or the given cycle time), weight, means of payment, etc. Figure 2.5 gives a structured overview of the dimensions of C&P.

1) Cf. e.g. Ellinger et al. (1980), Troßmann (1983).

2) Cf. Wäscher (1989b) p. 43.

3) Cf. Wee/Magazine (1982).

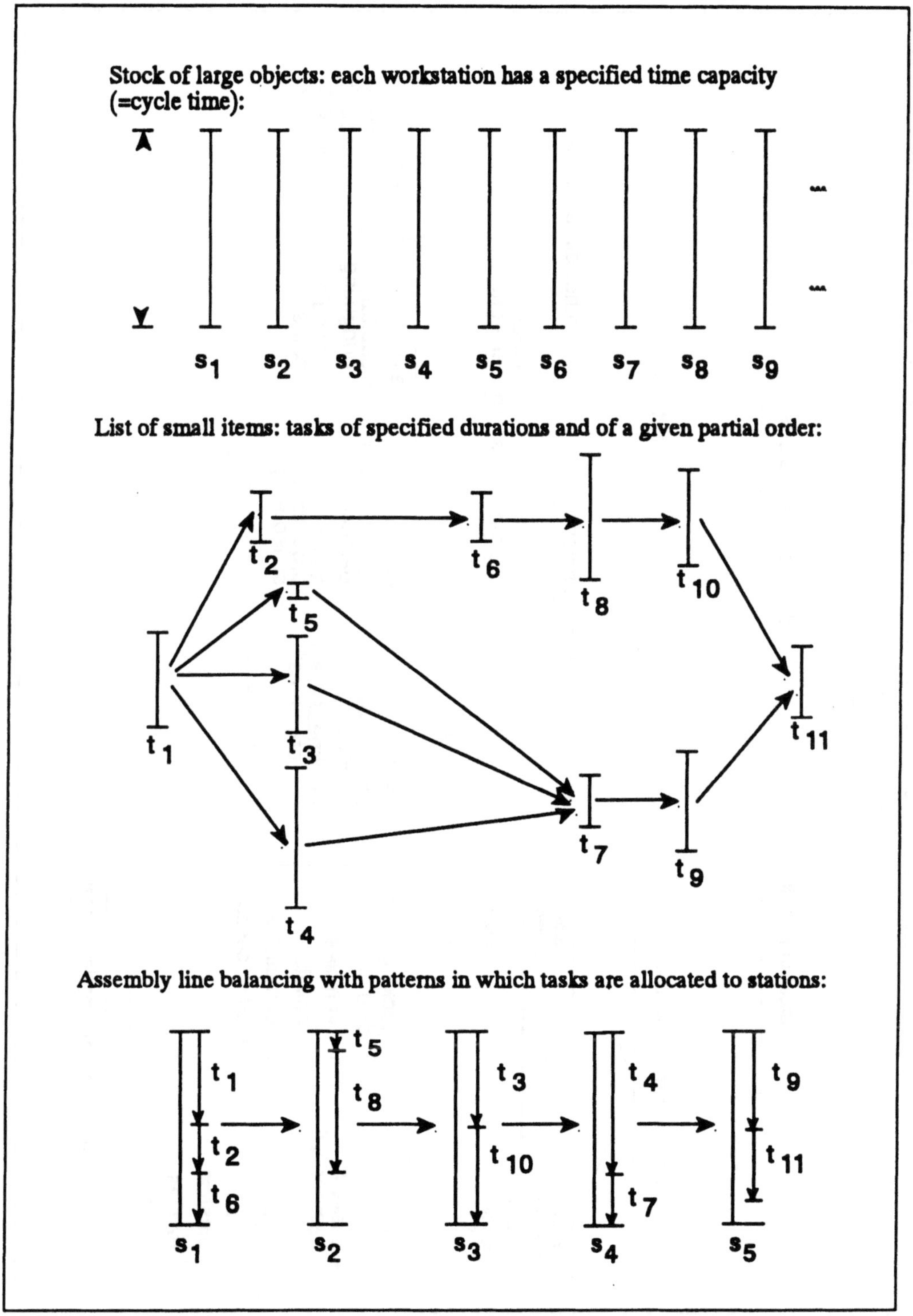

Figure 2.4: Assembly line balancing as an abstract C&P problem

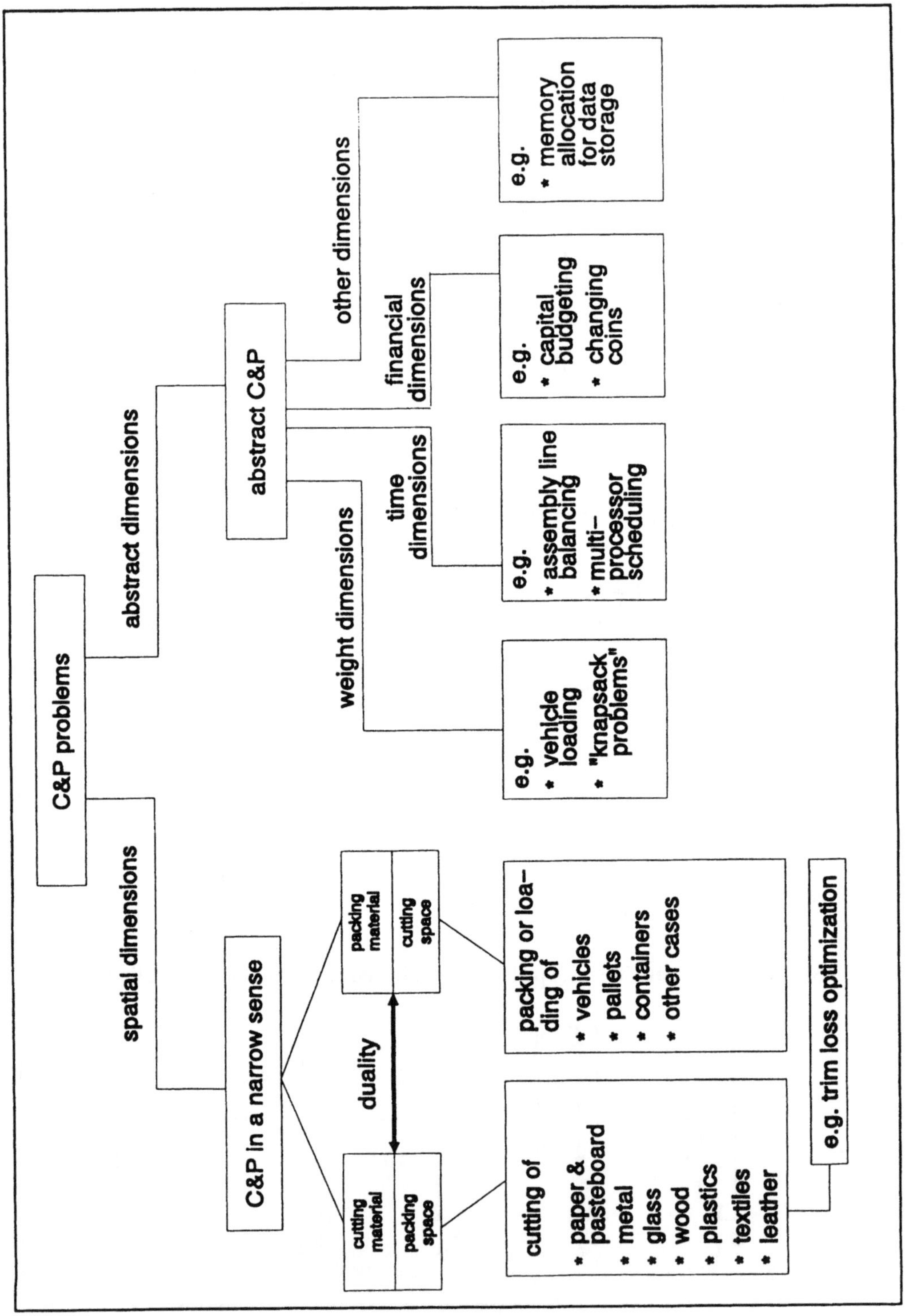

Figure 2.5: Phenomena of cutting and packing

2.2.1. Cutting and Packing in Spatial Dimensions[1]

Cutting and packing problems concerning material objects and items are in general defined within the three dimensions of the Euclidian space.

(1) Cutting problems (in a narrow sense)
In cutting problems (in a narrow sense), material objects of a given size are divided into smaller pieces (e.g. paper reels[2], iron slabs[3], corrugated paper[4], textile rolls[5], steel plates[6], and glass plates[7]). The problem typically appears in production processes when the available stock material does not have the necessary measurements for in-plant processing or to complete the customer's orders. In order to fulfil the cutting orders economically, it is usually important to minimize trim loss.

(2) Packing problems (in a narrow sense)
In packing problems (in a narrow sense), material items are assigned to a specific area or case. This is exemplified in manufacturing plants, where products must be loaded onto a pallet for distribution[8], or in trade or service industries, where the problems occur in loading containers[9], aeroplanes[10] or ships[11]. An essential goal here is to maximize use of space.

2.2.2. Cutting and Packing in Abstract Dimensions[12]

In a broader sense, cutting and packing may be concerned not only with the three dimensions of Euclidian space, but also with other dimensions.

---

1) Cf. Dyckhoff (1987) pp. 14 ff and (1990) p. 148.
2) E.g. Haessler (1971).
3) E.g. Tilanus/Gerhard (1976).
4) E.g. Moll (1985).
5) E.g. Gehring et al. (1979), Troßmann (1983).
6) E.g. Haessler/Vonderembse (1979).
7) E.g. Dyson/Gregory (1974).
8) E.g. Puls/Tachoco (1986).
9) E.g. Liu/Chen (1981).
10) E.g. Larsen/Mikkelsen (1980).
11) E.g. Haessler/Talbot (1990).
12) E.g. Dyckhoff (1987) pp. 20 ff and (1990) p. 148.

(1) Dimensions of weight

When the constraining factor in loading is the weight of cargo and not the spatial capacity, weight serves as the decisive factor. The cases must be packed in such a way as to maximize usage of loading capacity (large objects) with the given orders (small items). The unused capacity is considered as trim loss. Examples here are the classic knapsack problem[1] and vehicle loading problems.[2]

(2) Dimensions of time

C&P problems with respect to dimensions of time appear when tasks (small objects) are bound by time specifications which can be described in geometric terms. The sum of the durations of small tasks may not exceed the time capacity (large object). Unused time-capacity is considered trim loss. Examples are the above mentioned assembly line balancing, the scheduling of multi-processors[3] or the programming of commercials in an allotted time span[4].

(3) Financial dimensions

C&P problems with financial dimensions appear when for example a given amount of capital is to be distributed among several inseparable investment projects. Here, the aggregate value of the investment should be maximized (capital budgeting problem)[5] or, in banking, a given sum of money should be partitioned into a minimum number of banknotes or coins (problem of changing coins)[6].

(4) Other dimensions

A further aspect of abstract C&P problems handled in the literature is memory storage allocation[7], where pieces of information must be stored efficiently on a datamedium with a given capacity.

The above aspects of dimensionality have the common characteristic that they directly or indirectly concentrate on trim loss minimization by economically handling material or abstract matter. This is not necessarily the only objective relevant for C&P problems. Examples are in the locational planning of production facilities within the factory[8] and the

---

1) Cf. Dantzig (1957).
2) Cf. Eilon/Christofides (1971).
3) E.g. Coffman et al. (1978a).
4) E.g. Brown (1969), Brockhoff/Braun (1989).
5) E.g. Lorie/Savage (1955) pp. 232 ff.
6) E.g. Martello/Toth (1980a).
7) E.g. Garey/Johnson (1981).
8) Cf. Wäscher (1984) pp. 930 ff, Domschke/Drexl (1985).

construction of integrated switching circuits in a way which minimizes transport routes and costs, or pure layout problems in which perhaps 'only' packing stability is to be considered.

2.2.3. Related Problems

Additionally, there are a number of other problems which can be connected to geometric-combinatoric problems. An example is the so-called assortment-problem[1], which concerns the question of optimal dimensionalization of large objects or small items with respect to improving the use of resources. Another problem is that of packaging design.

2.3. Delimitation in Investigation

The purpose of this book is not to deal with the geometric-combinatoric aspect of C&P problems in general, but to focus on characteristic problems in spatial dimensions, in which the main goal, directly or indirectly, is to reduce trim loss. Nevertheless, Appendix II provides an overview of literature for abstract and related problems.[2] Many of the assertions concerning spatial dimensions serve also for abstract and related problems. More particularly, the properties developed in Chapter 4 for the characterization of the logical structure of C&P problems can be applied to all phenomena.

1) Cf. Chambers/Dyson (1976) and Hinxman (1980).

2) The sources on this topic were found at random during the search of sources for C&P problems in spatial dimensions. For this reason the synopsis given in Appendix I is not complete.

3. The Treatment of Cutting and Packing Problems in the Literature

3.1. Models as Idealized Images of Actual Phenomena

In order to successfully define C&P problem types it is necessary to undergo a structural investigation of actual problems and/or an analysis of the developed decision models. In this book we have chosen the latter approach, primarily for two reasons: first due to the multiplicity of phenomena it would be practically impossible to find a substantial empirical basis of actual problems through a data collection within the scope of this investigation. Second, using existing models as a basis for systematization harmonizes with the fundamental character of this book. It also speaks to the practical side of this investigation, to document and systematically describe the development of research in this field. This procedure does, however, present some problems. One important question is to what extent models in general and decision models in particular can be considered as images of actual phenomena.

As a rule, models for solving decision problems are developed through a 2-stage process. In the first stage a conceptual, usually verbal model ('real' model) of an actual problem ('real' problem) is formulated, which abstracts from the actual problem in that it concentrates on problem-relevant aspects. In the second stage the conceptual model is reduced to a workable mathematical formulation which then leads to suggestions for practical usage through mathematical procedures.[1] The difficulty here is that a model of an actual situation is neither independent of context nor will it be determined solely by the characteristics of the problem (see Figure 3.1).

A modelling process must be seen as a subjective, idealized operation, which is influenced by technical and methodological knowledge of the respective individual.[2] Thus, one's model-development process is based on a subjective "inner model" created out of an individual's environment, rather than on objective reality.[3] Technical knowledge particularly

1) Cf. Pfohl (1977) p. 27, Schneeweiß (1989) pp. 83.

2) See also the model concept by Gaitanides (1979) p. 16 and Bretzke (1980) pp. 23 ff.

3) Cf. Kirsch (1970) p. 76.

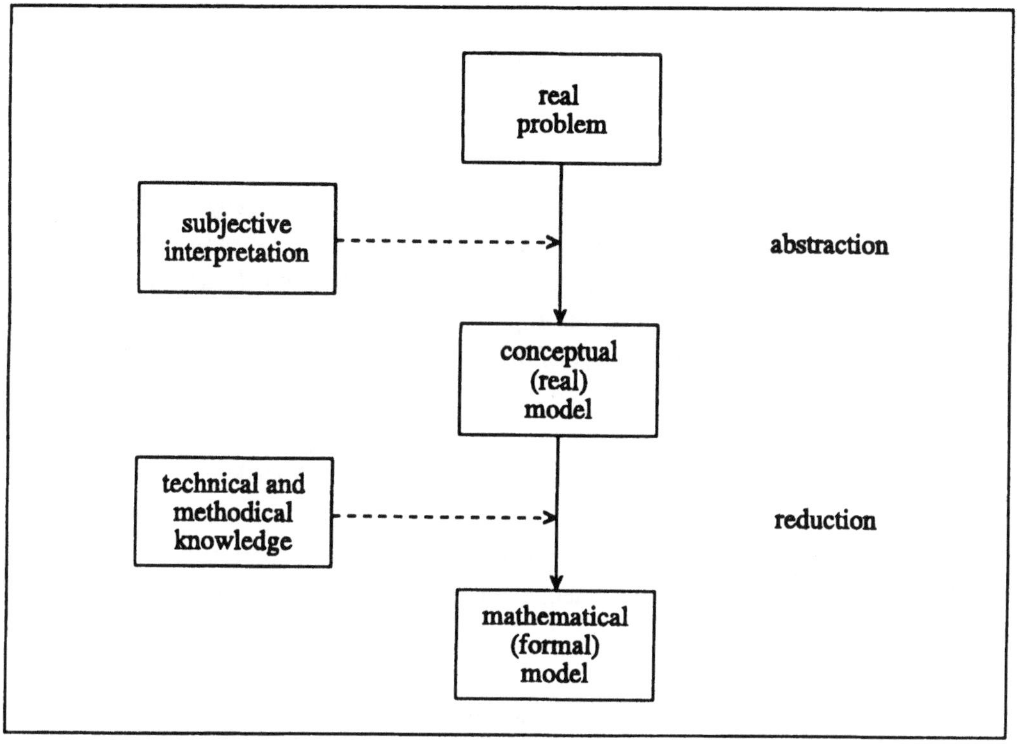

Figure 3.1: Model development of an actual problem

influences this "inner model" and therefore the manner of the model development for particular problems as well; a mathematical model without efficient solution methods lacks practical use and only serves for theoretical insight.[1] A 'model-builder' will formulate (simplify) a problem in such a way that it is efficiently solvable according to the background of his own technical knowledge. The result is that he often distorts reality, whether intentionally or unintentionally.[2] Because the degree of distortion is related to the level of abstraction in the process fewer distortions are to be found for decision models on the level of the 'real' (conceptual) models than for that of the 'formal' (mathematical) ones. Thus, by using (decision) models for the analysis of actual phenomena, it is appropriate to concentrate the investigation on the 'real' models.

In the literature on C&P problems the mapping of an actual phenomenon by a formal model and the mathematical deduction of solutions are basically the focus of interest.[3] The underlying 'real' model, which would be more useful with respect to the assessment of

1) Cf. also Müller-Merbach (1977) p. 18.
2) Cf. Lenz (1987) p. 276.
3) Cf. Pfohl (1977) p. 26.

phenomena, has often not been made explicit. Hence, it is usually difficult, if not completely impossible, to judge to what extent the subjective reduction of 'real' models into mathematical models is an appropriate simplification[1] or an inappropriate distortion[2] of actual given factors.

Accordingly, our analysis of the literature proceeded rather pragmatically. We accepted all models as 'reality' which were announced as such by the authors. Likewise, the developed types of actual C&P problems should be understood as idealized illustrations of actual phenomena, which are particularly useful in providing a deeper understanding the problem area and in serving as a starting point for further investigation.

3.2. Sources on Cutting and Packing Problems

The literature on C&P problems can be divided into various categories. We will start by differentiating above all according to thematic and bibliographical criteria.

3.2.1. Differentiation According to Topical Criteria

According to thematic criteria sources are divided into roughly three (possibly four) categories. The first category consists of case studies on actual, existing C&P problems. In the second category are theoretical/methodical studies, which deal with the development or improvement of various models and solution approaches and which are based on either abstractly formulated or previously published problems. The third category is a transition stage dealing with application-orientated studies which prove more related to practice than the theoretical studies, but are not represented in case studies. Existing overviews as a fourth category are a special type of contributions. These are primarily theoretical/methodical studies which describe solution approaches either for particular areas of usage or for various but more general and simplified problems.[3]

3.2.2. Differentiation According to Bibliographical Criteria

Sources for C&P problems are available in periodicals and textbooks for Management and Information Sciences, Engineering fields, Mathematics, Operations Research, etc., as well as conference proceedings and practitioner magazines (such as *Holzzentralblatt* and *Glas-*

1) E.g. reducing 3-dimensional Euclidian space to decision-relevant dimensions.

2) E.g. for purely mathematical reasons.

3) There are very few exceptions to this kind of overviews, e.g. Dyckhoff et al. (1985).

welt). C&P problems have also been analysed in unpublished habilitation theses, dissertations and masters' theses, as well as numerous other unpublished reports.

3.3. Delimitation of Investigated Literature

The book concentrates on sources which fulfill all the following conditions:

- they deal with C&P problems in spatial dimensions[1];
- they are case studies, application-oriented contributions or at least theoretical/methodical studies with a recognizable degree of practical relevance;
- and they are published in scientific journals or books in German or English.

There are three reasons for these restrictions:

The first reason is that these sources characteristically identify C&P problems which do in fact exist in practice and are described in sufficient detail. We have chosen not to analyse many of the purely mathematical sources which are found in the literature on bin packing or knapsack problems as they are of little relevance in the construction of C&P problem types. Of course they are of great interest to their solution. Hence, such publications are listed in Appendix I under the term 'Mathematical Problems'.

Second is that these sources have generally achieved a certain 'scientific standard'[2] as far as the quality of the models representing the actual problem is concerned. Furthermore, the chosen contributions are usually available to every scientist or interested expert[3] without great expense. This favours the further intention of this study, to create an easy access to existing knowledge.

Third it was necessary for practical reasons to impose deadlines and a limit on the number of publications dealt with. The deadline was the beginning of 1991. Some of the publications of 1990 or even 1989 may not have been available or even known to the authors by this date.[4]

---

1) See Chapter 2.3.

2) It should not be said that other individual sources can not provide sufficient information.

3) It was our experience, that it is often difficult, wearisome and at times impossible to obtain other kinds of literary sources, in particular the 'grey literature'.

4) Literature which was first discovered or available after the editing finished is in Part C of Appendix I.

Due to the delimitation of sources[1], 308 problems from 269 publications were analysed with reference to the definition of types.[2] Surveys from scientific journals and books were given special attention as they provide an introduction to the topic of Cutting and Packing. These studies were analysed according to the handling of fundamental problem types and are presented in Table A.1 of Appendix I.

1) Since it was not always possible to arrange the individual contributions into the given categories, some were classified arbitrarily.

2) Nearly 700 sources have been reviewed by the authors, not all of them are listed in the bibliography (e.g. old working papers).

4. Systematic Catalogue of Properties for the Characterization of Cutting and Packing Problems

4.1. Basis for Characteristic Properties

In order to achieve the goal of comparing C&P problems from the viewpoint of associated solution approaches, it is necessary to describe the numerous problems presented in the literature in a clear, systematic way. Simply giving a description of the actual phenomena would fail because of its multifarious nature. Only few forms could be studied, which would not suffice to make proper generalizations. Therefore it is necessary to develop a systematic instrument to condense the numerous actual problems to a few fundamental ones, which then allow for conclusions for any amount of practical variants.[1]

In this study, typology is the scientific method used for systematic condensation of actual problems.[2] In this process attributes are first identified representing the various practical variants of C&P problems which are then differentiated into relevant properties. These characteristics, i.e. attributes and their properties, are developed into a systematic cata-

1) Cf. Große-Oetringhaus (1974) p. 20 and p. 338.

2) Typology is understood as the entirety of all thought processes and their outcomes, which bring order to actual phenomena in a field with respect to the pursued goals of investigation. The differentiation of attributes, which characterize the field of study, and the arrangement of properties into meaningful phenomena lead to the formation of types and eventually to type-series. The first depict the elements of order and the latter the groups of order. Cf. Große-Oetringhaus (1974) pp. 26 ff.

logue.[1] On the basis of this catalogue one can advance to the second level, which is to combine characteristics into various types of C&P problems.[2]

4.2. Design of the Catalogue

In conceptualizing the catalogue it is reasonable to distinguish two ways in which the problems differ:

- First, one must show how problems can be characterized independently of their physical mode of application. That is, characteristics should be determined which closely describe the various manifestations of the fundamental logical structure of C&P problems (**characteristics based on the logical structure**).

- Second, one must examine how C&P problems vary according to their practical background. That is, the surrounding factors of the problem must be examined (**reality-based characteristics**).

The following method has been employed to select characteristics:

The identification of the main attributes in Table 4.1 attempts to operationalize essential characteristics which must be considered in the development, selection and application of solution procedures. Each of these main attributes is further differentiated by specification of properties, which are then further detailed so that they themselves represent attributes.[3]

No difficulties arise in presenting characteristics based on the logical structure, since they are revealed in the basic logical structure of C&P problems explained in Section 2.1. In fact, differences between C&P problems result from various properties of the main attributes

1) Typology should not be confused with classification, which also organizes elements and their relation to each other, but which, through its completeness and clarity, serves a different goal. The most famous example of classification is the periodical element system. From the knowledge of its totality it was possible for the discoverers Mendelejev and Meyer to determine elements not yet known, thus relating theory and reality. In typology this connection is neither possible nor desired, as exactly the opposite reasoning process is employed, in which a type is based on experience in reality. It is possible, however, to build a typology upon a classification scheme. It is not necessarily possible to form a type, however, even if the totality of a process is known. First the actual existence must be tested; cf. Große-Oetringhaus (1974) p. 40.

2) Cf. Chapter 5 to 9.

3) Appendix II gives a brief description of all characteristics.

listed in column 2 of Table 4.1. Section 4.3. goes into more detail concerning those main attributes.[1]

categories	attributes based on the logical structure	reality-based attributes
main attributes	• dimensionality • type of assignment • characteristics of large objects • characteristics of small items • pattern restrictions • objectives • status of information and variability of data • solution methods	• kind of objects and items • branch of industry • planning context • software

Table 4.1: System of main attributes

Presenting reality-based properties leads to difficulties, because in practice, so many varying characteristics exist. It would thus be necessary to use a large number of properties to cater for every actual situation. This, however, would be neither useful nor necessary. The practical background of a C&P process can be recognized when:[2]

- the contextual meaning of abstract properties based on the logical structure is clarified through providing an actual label of the objects and items,

- the sphere of actual experience in which the problem is rooted is explained, and

- the interdependence to other areas of planning and function is demonstrated.

1) Treating solution methods as attributes based on logical structure is caused by the idea that they can serve in the assessment of properties of other attributes; see Section 4.3.7.

2) Cf. Dyckhoff (1987) p. 171.

Thus the reality-based main attributes listed in column 3 of Table 4.1 can be identified. They are presented more thoroughly in Section 4.4.[1]

4.3. Characteristics Based on the Logical Structure[2]

4.3.1. Dimensionality

Unquestionably, the most vital characteristic in C&P problems is dimensionality. In the physical world, (material) objects and items can be described by the three spatial dimensions. But in cutting and packing this spatial dimensionality is not important. Only relevant is the minimum number of dimensions needed to geometrically describe a cutting or packing pattern.

Thus the following dimensions appear:

- one-dimensional,
- two-dimensional,
- three-dimensional, and
- multi-dimensional problems (more than three dimensions).

In one-dimensional problems objects and items are defined by their length. An example is cutting rods or tubes, where an object of a given diameter is to be divided into shorter parts.

In two-dimensional problems, the objects and items are surfaces. Flat materials (e.g. sheet metal or glass plates) must be cut into smaller sizes of the same material thickness.

There are relatively few actual situations in which a problem arises affecting all three spatial dimensions. They occur most often while loading containers, aeroplanes or ships. But even here one or more of the dimensions often can be standardized, so as to reduce the complexity of the problem to fewer dimensions.

Multi-dimensional problems occur especially in abstract C&P problems (e.g. in the case of multi-period capital budgeting)[3]. They can also occur in loading containers, however, when

1) The assignment of software to reality-based attributes is purely arbitrary for the time being. As is demonstrated later, this is because all reality-based characteristics, and thus the software, could only be analysed with application-orientated papers and case studies.

2) Cf. Dyckhoff (1990) pp. 150 ff.

3) E.g. Lorie/Savage (1955).

an item is to be stored within a certain time frame, and time is considered as the fourth relevant dimension.

4.3.2. Type of Assignment

In the C&P process a suitable assignment of items to objects is sought which satisfies the combinatoric base condition, that is, that each large object is assigned a given set, possibly empty, of small items and each used small items is assigned to exactly one large object. The basic differences in the structure of C&P problems result from whether a complete stock of objects or items must be assigned or only a suitable selection. As a result, basically four cases appear:

- Type I : all objects, all items,
- Type II : all objects, selection of items,
- Type III : selection of objects, all items,
- Type IV : selection of objects, selection of items.

(1) Type I (Pure layout problems)
This situation arises when all of the objects and items must be assigned. This case is particularly important when the spatial arrangement of small objects is the essential question, as in machinery location in factories. This is a pure layout problem and is considered part of every C&P problem (cf. Section 2.1.).

(2) Type II ('Beladeprobleme')
To each large object at least one small object has to be assigned.

(3) Type III ('Verladeprobleme')
This is practically the opposite to Type II in that all small items have to be assigned to a selection of large objects.

(4) Type IV ('Ladeprobleme')
This occurs when the selection applies to both objects and items, so that unused items as well as objects can remain.

4.3.3. Characteristics of Large Objects and Small Items[1]

All C&P problems are similar in that the objects and items can be represented as specifically sized geometric bodies in a correspondingly dimensioned space of real numbers. The differences result from various properties of the following attributes:

- type of quantity measurement,
- figure (shape),
- assortment, and
- availability.

(1) Type of quantity measurement

The number of objects or items in C&P problems is normally measured discretely, i.e. by natural numbers. There are other problems, however, in which one can find variable measurements of objects or items in relation to a relevant dimension. A typical example is in cutting rolls of a given material, when it is up to the manufacturer to decide how many rolls or parts of rolls must be used to obtain the desired length of the order. In this sort of problem with actually two relevant dimensions, only one of which is fixed, the material is not measured according to the number of rolls but in continuously variable measures of length (see Figure 4.1). Thus, a problem which initially appears to be two-dimensional is reduced to one dimension with continuous quantity measurement of the objects or the items, respectively. The lengths of the rolls become fused into 'super rolls' and the length no longer serves as a relevant dimension.[2]

(2) Figure (Shape)

The figure (or shape) of an object or item is defined by its form, size and orientation within the relevant dimensions. When all of these three qualifications coincide in comparing objects or items they can be considered identical.

Two objects or items have the same form when they would be congruent after changing their size proportionally with respect to all relevant dimensions. In one-dimensional space the only possible form is a line, which means all objects have the same form. For two- or three-dimensional objects or items, however, there are numerous possible forms. Particularly significant to the solution of a C&P layout problem is to determine whether the objects are only rectangular (rectangles or blocks) or have non-rectangular forms (e.g. triangular, circular or irregularly-shaped) as well.

---

1) Objects and items are characterized by essentially the same characteristics and are dealt with simultaneously in this section. In concrete cases the properties are not necessarily identical and therefore must be handled separately.

2) One-dimensional problems with continuous measurements in length are often called 1.5-dimensional problems; cf. Dyckhoff et al. (1984) pp. 915.

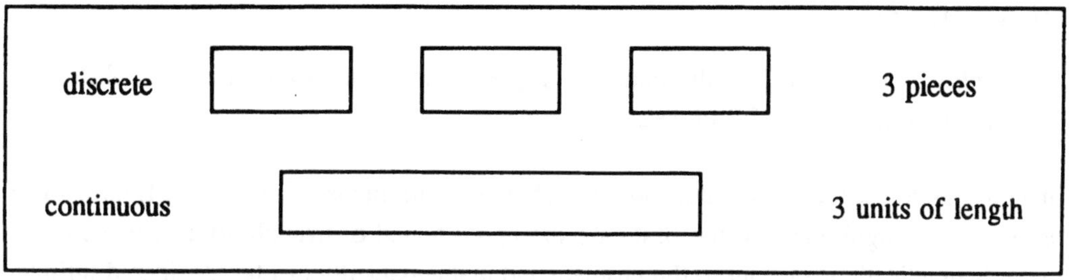

Figure 4.1: Discrete and continuous quantity measurement

The size of an object or item results from its measurements (e.g. length, width, weight etc.) in the relevant dimensions. In this way, squares may have the same form, but their figures differ according to their size. An important point in reference to the difficulty of solving C&P problems is the relation in size between large objects and small items.[1] Therefore differences in characteristics are derived from the relative sizes of objects and items.

The orientation of an object or item is determined by its position in a given space. The differences in characteristics depend on whether they have a fixed orientation or a variable one. If rotation is possible, two- and three-dimensional problems differ as to whether there are allowed any rotation, a 90-degree turn parallel to a given surface or any 90-degree turn.

(3) Assortment

Assortment can be defined as the entire group of existing figures of the objects and items, respectively, differing between homogeneous or heterogeneous groups.

In homogeneous assortments all objects or all items, respectively, are congruent (i.e. possess the same form and size), possibly varying in their orientation. In special cases, when not only form and size but also orientation are equal, the objects or items have the same figure, i.e. they are identical. The figure can either be predetermined or, to a certain extent, variable as in the 'overpacking' of pallets.

With heterogeneous assortments the objects or items, respectively, may have different measurements in the relevant dimensions, which can either be predetermined or variable. This case must then be further classified as to whether many different figures exist or whether enough items of identical figure exist so that they can be divided into representative groups according to their figure.

1) Cf. e.g. Haessler/Talbot (1990).

(4) Availability

Availability is understood as the quantity, sequence and date, with which the objects or items correspond to the C&P process.

In consideration of the available number of objects or items of the same figure, three properties are significant. In the first case the upper bound of availability can be so large, that it is practically irrelevant to the problem and mathematically can be considered infinite. The second case presents the opposite effect, where there is more or less a limited amount of available objects and items, respectively, or in extreme cases only one. The third case is a combination of the previous two, where the amount of some of the figures is limited and others unlimited. This case often appears in cutting processes where residual pieces of previous orders are stored for future use.

In most cases, no relationship exists between the objects or items themselves, although it is possible that the order in which they are to be cut or packed is prescribed to some extent. In a partial order one object or item should be placed before another, as with small items on an assembly line; a complete order is one where there is a given sequence for the entire group of objects or items, as in cutting red-hot slabs of pig-iron.

The date indicates whether the objects or items have identical or different time references (e.g. varying delivery dates).

4.3.4. Pattern Restrictions

Pattern restrictions refer to conditions for the assignment and spatial arrangement of small items within the larger object. Aside from the combinatoric and geometric base conditions there are often further technical and organizational restrictions as well. They can then be classified as to whether they are pattern restrictions in a narrower sense, i.e. geometric contraints with regard to the design of a pattern, or whether they refer to the assignment of patterns among one another (operational characteristics) or to the allocation dynamics.

(1) Geometric characteristics

Geometric characteristics in C&P problems can result from technological restrictions as well as organizational conditions, which must be observed in pattern design. The following three categories of restrictions can be differentiated:

- distancing restrictions,
- frequency limitations, and
- separation restrictions.

Distancing restrictions occur more often in cutting than in packing problems, for example when a certain distance must be kept, either between the items to be cut or to the edge of large objects. This is relevant for tasks such as avoiding shifting in glass-breaking or removing defects of the edges in cutting paper.

Frequency limitations relate to the number of items and to the number or combination possibilities of figures in a pattern. It may be necessary for organizational reasons to combine each customer's order into one pattern, or to limit the number of items in a pattern with regard to technological possibilities of a cutting machine.

Separation restrictions refer to the type of arrangement for small items in a pattern. They are particularly of practical importance with two- and three-dimensional problems. In fact, the investigation can be reduced to the theoretically and practically dominant case of rectangular units. Fundamentally, rectangular units can be cut or packed in

- orthogonal or
- non-orthogonal patterns.

In orthogonal patterns small items are arranged parallel to the large object's edges, whereas in non-orthogonal patterns the angle is optional.[1] In the former either a guillotine or a nested pattern can be employed (Figure 4.2). Cutting problems normally employ a guillotine pattern, as its technology characteristically allows only uninterrupted cuts from one end of the object to the other. Often the number of steps in possible guillotine cuts is limited. On the other hand, interrupted cuts are allowed at any point in a nested pattern. This pattern is especially useful in packing, for instance to avoid problems of instability in pallet loading.

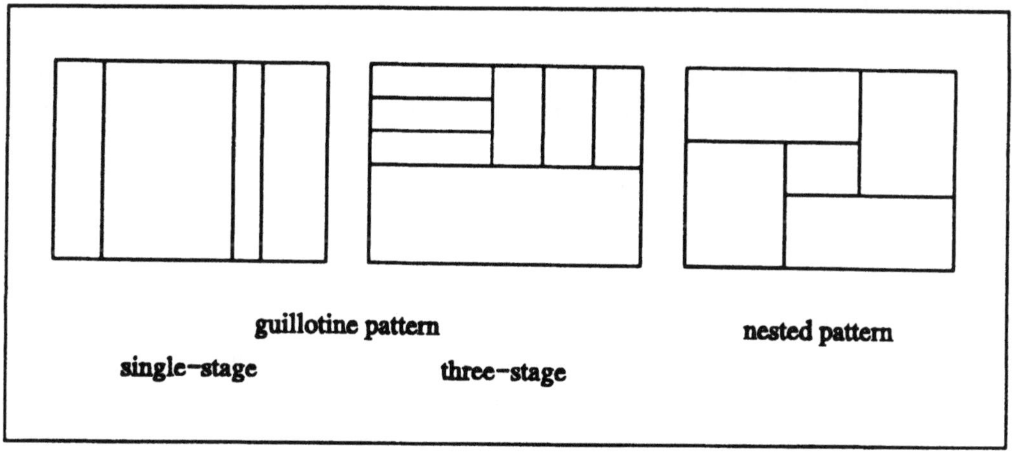

Figure 4.2: Examples of orthogonal patterns in C&P problems

1) Non-orthogonal patterns naturally include all patterns for irregularly-shaped items.

(2) Operational characteristics ('organization')

The systematization of operational characteristics in C&P problems is based on the following attributes:

- number of stages,
- pattern sequence, and
- pattern frequency.

A stage in a C&P process is defined by each independent work cycle of cutting or packing. In single-stage processes the task is performed in a single, continuous operation, with minimal interruptions. This means that also processes which actually are carried out by several successive operations (e.g. in dividing objects by guillotine cuts or in loading containers) are treated as if all the operations were performed simultaneously in one step. The only relevant point is the result of the cutting or packing process. C&P problems are only considered multi-staged when the process has at least some interruptions timewise, as in the paper industry where rolls of paper are often hung up and layed aside until being cut further. In this case several production units are in use which should be considered when designing a cutting pattern. One must also take into account the capacity for buffer stores in multi-stage patterns.

Often, the pattern sequence in the C&P process is determined in a separate planning stage rather than being considered part of a problem's logical structure. There are, however, several factors which require the consideration of pattern sequence while actually developing a cutting or packing pattern. For example, due to organizational reasons it may be necessary to process various orders of a certain customer directly one after the other.

Restrictions in terms of pattern frequency are typically found in cutting problems. In order to reduce the frequency of pattern alterations and therefore the cost of change-over in cutting facilities, there are often given minimal numbers for the application of the same pattern, or a limit for applying different patterns.

(3) Allocation dynamics

C&P problems are generally treated as static problems, that is, all objects and items are present in the same period of time. In reality, however, there are situations in which objects and items, respectively, may have various deadlines. It is possible, for example in aeroplane or container loading that a number of small items must be unloaded before others. A packing pattern is thus developed in such a way as to avoid unnecessary repacking.

Static problems also differ according to whether all objects are identified at the time of planning (off-line problems) or only partially identified (on-line problems).

4.3.5. Objectives

The solutions to C&P problems must be feasible, fulfilling combinatoric and geometric conditions, as well as possible technical and organizational restrictions. Mapping of a C&P problem in a useful decision-model also requires the selection of criteria and the formulation of objective functions.

There are several objectives relevant to C&P planning. Within an organizational framework, the planner is given the objectives by the management of the firm. By redefining these main goals into operational subgoals, a catalogue of competing goals is often developed, which can only seldomly be integrated into a single utility function. Rather, interactive approaches of multicriteria decision making are suitable which however have little application in practice until now. Usually, only one objective is maximized or minimized whereas the other objectives - if at all - appear merely as levels of satisfaction, control parameters or fixed values in the search for a solution.

Due to the multitude of actual C&P problems and their background in the practical realm it is not possible to develop a criteria system suitable for all situations. There are, however, several particularly characteristic objectives:[1]

- Input minimization: minimizing the number of large objects used

- Trim loss minimization (absolute): minimizing trim loss quantity

- Trim loss minimization (relative): minimizing trim loss quantity in relation to input quantity

- (Materials/space) cost minimization: minimizing object usage or trim loss according to cost per piece, depending on its figure

- Change-over cost minimization: minimizing costs for altering cutting or packing facilities

- Inventory cost minimization: minimizing the stock of large objects and small items

- Value maximization: maximizing the value of packed or cut small items

1) A comprehensive discussion of relevant objectives for a particular type of cutting problems can be found in Wäscher (1989b) pp. 93 ff.

4.3.6. Status of Information and Variability of Data

Due to their usually fairly short planning horizon, data in C&P problems are generally determined. There are cases, however, in which the data (for example the exact measurements of the objects or items) is of more a stochastic nature. An example is in rolling thin plates from hot pig-iron slabs, where, for technical reasons, the slab measurement varies to an extent and therefore cannot be exactly determined during planning.

Generally, order specifications determine the type and quantity of the items to be cut or packed. A certain degree of data variability can be allowed, for example in the paper industry, in which up to five per cent deviation in the measurements is often accepted upon delivery.

4.3.7. Solution Methods

Since our investigation is based on already formulated decision models rather than on actual C&P situations, it is often quite difficult to identify the actual underlying problem (see Section 3.1.). As an example, with existing model formulations it is nearly impossible to decide whether there are few or many different types of small items in a heterogeneous assortment. It is possible, however, to draw conclusions from the respective solution methods, which can be divided into roughly two categories:

- object-orientated methods and
- pattern-orientated methods.

In object-orientated methods (e.g. Next-Fit-Decreasing) the small items are individually assigned to the large objects. These methods are most applicable when there are many small, variously-shaped items.

On the contrary, pattern-orientated methods involve constructing diverse feasible patterns and then choosing the 'best' ones for the assignment. This category varies between those methods which produce one pattern after another and immediately select an appropriate assignment of objects and items (single-pattern methods) and those which develop several patterns at the same time and by simultaneous assignment can take into account possible interactions between patterns (multiple-pattern methods).[1]

---

1) The classification of solution methods into object-orientated, single-pattern and multiple-pattern methods is based on Sweeney's bibliography (1989) of C&P problems. He classifies sources according to these three categories and their dimensionality. Another contribution is by Dyckhoff (1990), in which the same terms are used, although expanded to other contexts, particularly those relating to single-pattern techniques. Since the attribute 'solution methods' is more of a supplementary nature in this investigation and the results of this study are not dependent on its classification, Sweeney's interpretation has been employed in order to make the results comparable.

4.4. Reality-Based Characteristics

Up to this point we have examined those properties of C&P problems which characterize the logical structure of the problem. In this section further characteristics will be introduced which lead to conclusions based on the actual background of a problem. Actual cases often contain important aspects which are simply too complicated to formulate mathematically, but which can be better recognized by a closer examination of the practical background of the problem.

4.4.1. Kind of Objects and Items, and Branch of Industry

It is normally feasible to identify the practical background of a problem by names of large objects and small items and the industrial sector concerned. For example, cutting tree trunks can serve for the production of wooden slabs or blocks, whereas pallet loading refers to industrial firms or the service industry.

4.4.2. Planning Context

As a rule, C&P problems are embedded in wider planning contexts. As the various planning areas are often interdependent, it would be of general interest to define the problem in such a way that as few as possible of the related dependencies are lost. In characterizing the planning process, it is reasonable to handle cutting and packing problems separately due to the fact that their problem setting is very different .

Cutting as a part of production planning reveals particularly strong interdependencies with other areas of planning, such as:

- order management,
- timing,
- capacity planning,
- sequencing,
- choice of manufacturing method,
- numerical (machine) control,
- inventory planning,
- material requirements planning,
- procurement planning.

Ideally, all of these areas are planned simultaneously, although this is presently impossible due to current (possibly also future) means of data gathering and processing. In order to be able to consider these interdependencies, two procedures are generally implemented:

- All areas are planned successively, where solutions of previous problems provide restrictions for further planning. In this procedure the cutting process is planned separately.

- Two or more areas of planning are integrated into a simultaneous planning process, while the other areas are planned successively.

In packing problems the relations to other planning areas are not as obvious as in cutting problems, which means that an isolated packing plan is considered the most appropriate one. Such strong dependent factors to other areas of planning which require simultaneous planning occur in only a few cases, namely the interdependency with

- balancing and
- stability considerations,
- packaging planning,
- transport planning.

4.4.3. Software

The characteristic 'software' has a special position among the main attributes. It does not belong to the definition of a type, and yet has an important role in assessing the applicability of decision models found in literature. Software can be developed either for practical application or exclusively for purposes of simulation (testing algorithms). Decision-models based on the former software are certainly closer to reality than those of the latter. Furthermore testing this attribute also serves to identify types of C&P problems for which "practically tested" solution concepts already exist.

4.5. Overview

The isolated presentation of characteristics of C&P problems does not mean that they are independent of each other. There are, in fact, close interactions, sometimes conflicting, where the characteristics can only be assessed in relation to the process as a whole. An overview of all the attributes and their properties can be found in Table 4.2 (characteristics based on the logical structure) and Table 4.3 (reality-based characteristics).

attributes	properties	
dimensionality	- one-dimensional - two-dimensional	- three-dimensional - multi-dimensional
type of assignment	- Type I (all objects, all items) - Type II (all objects, selection of items) - Type III (selection of objects, all items) - Type IV (selection of objects, selection of items)	
characteristics of small items and large objects	**(1) type of quantity measurement** - discrete - continuous **(2) figure** (shape) - form -- rectangular -- non-rectangular size - orientation -- fixed -- optional -- 90°-turns (parallel) -- 90°-turns (optional)	**(3) assortment** - homogeneous (extreme: identical) -- fixed -- variable - heterogeneous -- fixed -- few objects per figure -- variable -- many objects per figure **(4) availability** - number per figure -- limited (extreme: 1) -- unlimited -- mixed - sequence -- no order -- partial order -- complete order - dates -- identical time references -- different time references
pattern restrictions	**(1) geometric characteristics** - distancing restrictions -- between items -- to the edges - frequency limitations -- limited combinations of figures -- limited number of figures -- limited number of items - separation restrictions -- non-orthogonal patterns -- nested patterns -- guillotine patterns with fixed number of stages -- guillotine patterns with unlimited number of stages	**(2) operational characteristics (organization)** - number of stages -- single-stage -- multi-stage - pattern sequence -- optional -- restricted - pattern frequency -- limited number of different patterns -- minimum number of identical patterns **(3) dynamics of allocation** - dynamic - static -- off-line -- on-line
objectives	- input minimization - trim loss minimization (absolute) - trim loss minimization (relative) - value maximization	- (material/space) cost minimization - change-over cost minimizaion - inventory cost minimization - others
status of information/ variability	**(1) status of information** - deterministic - stochastic - uncertain	**(2) variability of data** - fixed data - variable data
solution methods	- object-orientated methods	- pattern-orientated methods -- single-patterned -- multi-patterned

Table 4.2: Characteristics based on the logical structure

<table>
<tr><th>attributes</th><th colspan="2">properties</th></tr>
<tr><td>kind of objects and items and branch of industry</td><td colspan="2">- kind of large objects
- kind of small items
- branch of industry</td></tr>
<tr><td>planning context</td><td>(1) cutting problems
- order management
- timing
- capacity planning
- sequencing
- choice of manufacturing method
- numerical control
- inventory planning
- material requirement planning
- procurement planning
- others</td><td>(2) packing problems
- balancing considerations
- stability considerations
- packaging planning
- transport planning
- others</td></tr>
<tr><td>software</td><td colspan="2">- practical application
- simulation</td></tr>
</table>

Table 4.3: Reality-based characteristics

5. Types of Cutting and Packing Problems in the Literature

In the previous chapter we have presented a catalogue of attributes and properties, which makes available a variety of possible characterizations of actual problems in cutting and packing. These are used in the following to define particular types of C&P problems.

5.1. Principles of Type Definition

A type is an idealized phenomenon which serves to identify the essential characteristics of the various forms of problems in an area of investigation. The definition of types is an intuitive, constructive process of abstraction, which must be both logical and factually true, that is, it must be based on both experience and empirical evidence. The decision which kinds and quantities of properties are necessary to characterize the essentiality of an actual phenomenon depends on the purpose a typology should serve.[1]

In the framework of this investigation, types of C&P problems are looked at as tools for the selection, development and application of solution procedures. Respectively, the purpose is to cluster those properties and to assign them to types in such a way as to clarify characteristics for the solution of a group of C&P problems. The goal is therefore not to simply apply all possible properties in the catalogue, but to combine into groups only those characteristics which are particularly relevant. These are termed **type-defining characteristics**.[2]

In defining types it is important to consider how they fit together to form a coherent whole, rather than treating each one in isolation. The types formulated in this investigation were not constructed as isolated units, but in regard to the overall context.

The results of the empirical investigation led to a hierarchical typology of C&P problems. First, **general types** are characterized by a few properties. These deal with types which are

---

1) Cf. Große-Oetringhaus (1974) pp. 26 ff.

2) Cf. Große-Oetringhaus (1974) p. 35.

adequately represented by the fewest possible characteristics.[1] These general types are then gradually differentiated into **special types** of possibly varying degrees of abstraction by further properties. The result is the development of systematized type-chains, in which each type describes an actual C&P problem with a certain degree of abstraction. The differentiation intensity and therefore the length of the chain depends on to what extent the examined literature handles a general type. The type catalogue does not guarantee its completeness, but is rather to be recognized as an order system based on the analysed literature and may have to be supplemented and expanded upon, for example by including abstract C&P problems.

5.2. Hierarchical Catalogue of Types

In order to empirically derive types for C&P problems, 308 problem situations were analysed out of 269 contributions up to the deadline in 1991. By "intuition and construction"[2], in a constant fluctuation between observation and mental classification on the one hand and consideration of plausibility and testing of validity on the other, 26 types of C&P problems were identified. These types will be presented in the following sections of this chapter.

5.2.1. General Types

On the first hierarchical level the following type-defining characteristics are identified according to the results of the empirical study, in reference to the construction of general types,

- 'dimensionality' with three properties: one-, two-, and three-dimensional
- 'type of assignment' with the properties: Type II through Type IV
- 'assortment of small items' with the properties: homogeneous, heterogeneous with few items per figure, heterogeneous with many items per figure.

'Assignment Type I' as a pure layout problem type has not proved to be of relevance for the C&P problems considered here.

The significance of dimensionality is revealed in the geometric complexity arising from the number of relevant dimensions of the respective mathematical problem. In one-dimensional problems the geometric aspect is trivial because such attributes as form and orientation of

1) Große-Oetringhaus (1974, pp. 31 and 61) defines this sort of type as an elementary type. Since this term includes only those characterized by a single property, it will not be used in this book.

2) Cf. Große-Oetringhaus (1974) p. 34.

the small items and separation restrictions are not of importance. On the contrary, those attributes unimportant in one-dimensional problems involve high degrees of complexity in more dimensional problems where spatial arrangement in three-dimensional problems is considerably more complex than in two-dimensional problems.[1]

The attributes 'type of assignment' and 'assortment of small items' first acquire a type-defining character by a suitable combination of their properties. As it is shown in Figure 5.1 four combined types of practical relevance can be identified. Their significance results from two aspects: First, the required solution methods are basically different. Second, all analysed sources can be included.

The four types are:

- bin packing type (BP)
- cutting stock type (CS)
- knapsack type (KS)
- pallet loading type (PL)

A **bin packing type** identifies a C&P problem in which numerous small, heterogeneously-shaped items are to be either completely (Type III) or incompletely (Type IV) assigned to a selection of objects. The problem of assignment in this situation is generally complex. This type of problem is NP-complete and is therefore most efficiently solvable using fast, simple heuristics (so-called BP-algorithms like Next-Fit, First-Fit etc.).[2]

A **cutting stock type** refers to those C&P problems in which many items of only several distinct figures must be assigned to a choice of objects, either completely (Type III) or incompletely (Type IV). As opposed to bin packing problems, the items can be divided into relatively few groups of identical figures (for the items of each group). As a result of this, there are often repetitive solutions for the layout of small items in large objects. Rather than considering each assignment individually, one layout can represent - or become a pattern - for numerous identical assignments. This, in contrast to bin packing problems, reduces the complexity of the problem to such an extent that it is often possible to use an exact procedure (e.g. simplex-algorithms) by omitting integrality conditions.

In a **knapsack type**, a selection from a large supply of items of heterogeneous figures must be assigned to only a limited stock of objects. Since all available objects are to be included, a problem in their selection does not exist. Branch-and-bound or dynamic programming procedures often serve as solution methods for knapsack problems.

1) Cf. Dyckhoff (1987) p. 194.

2) Cf. Coffman et al. (1984).

In a **pallet loading type** a selection of homogeneous small items must be assigned to large objects. Since the assortment of the large objects in this case is also generally homogeneous, one representative solution suffices for the task.

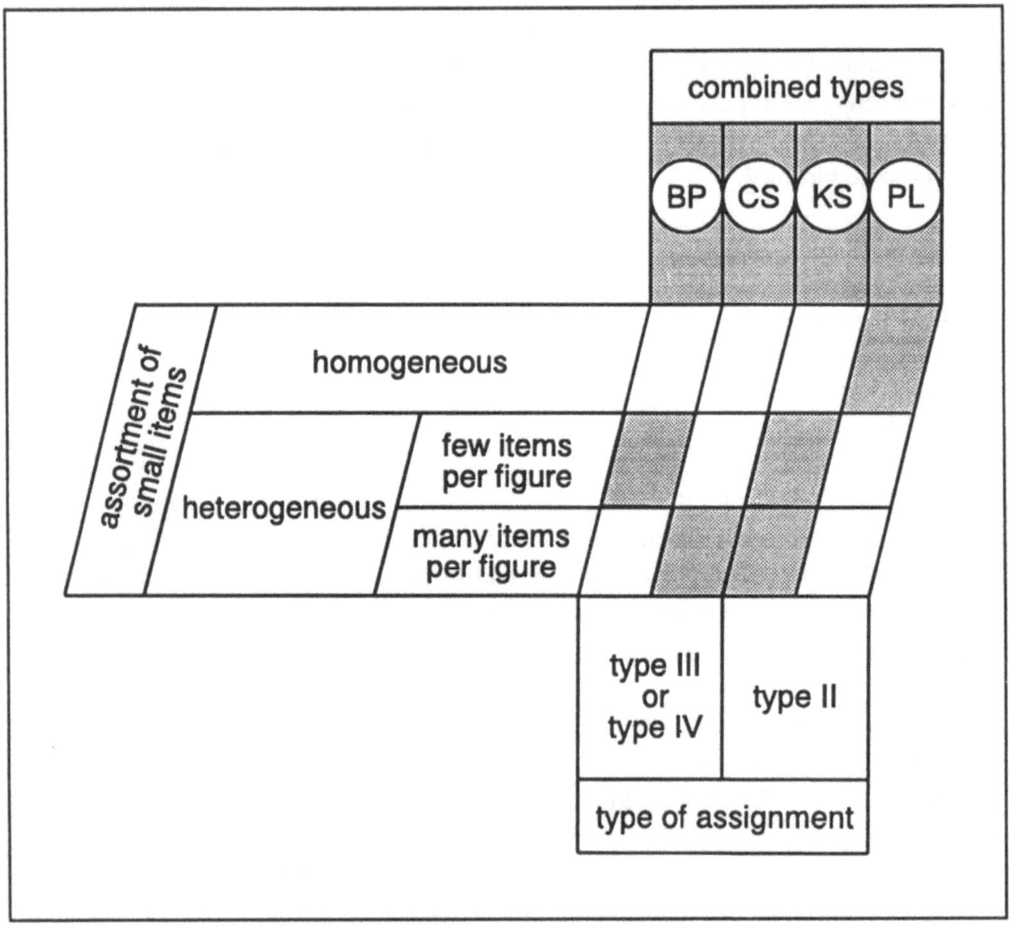

Figure 5.1: Combined types of the attributes 'type of assignment' and 'assortment of small items'

The names of the four types[1] have been chosen partly according to the area of application described in the literature. It may be pointed out, however, that our study defines 'bin packing' and 'knapsack' as broader, more general concepts as they are normally used in the literature.

Although the four types are of great significance when solving spatially dimensional C&P problems, in our view they adopt characteristics of general types only by taking into account their dimensionality. There are, substantial differences between one-, two- and three-dimensional problems in the four types.

Combining the level of dimensionality with the four given types results in 12 general types.[2] As is presented in Table 5.1, the following observations can be made with respect to their frequency in the analysed literature (308 problem situations):

dimensionality	combined types				
	BP	CS	KS	PL	
one-dimensional (1)	BP1 *47*	CS1 *96*	KS1 *9*	PL1 —	*152*
two-dimensional (2)	BP2 *31*	CS2 *70*	KS2 *21*	PL2 *20*	*142*
three-dimensional (3)	BP3 *0*	CS3 *5*	KS3 *7*	PL3 *2*	*14*
	78	*171*	*37*	*22*	*308*

Table 5.1: General types of C&P problems in the analysed literature

1) Wäscher (1989b, pp. 55) defines these types in a narrower sense, because he includes 'number of objects' as a supplementary property. In Dyckhoff (1987, pp. 202 ff) four types are included under the attribute 'assortment of items', namely bin packing, cutting stock, pallet loading, and vehicle loading problems. The first three types correspond to those stated here, the vehicle loading problem is not, however, identical to the knapsack type, but serves as an intermediate type between the bin packing and pallet loading types. Since vehicle loading problems have not led to independent solution approaches they are not considered here.

2) The one-dimensional pallet loading type is trivial and therefore not included in the following study.

- Three-dimensional bin packing types were not discussed in the analysed literature.[1]

- One- and two-dimensional problems in cutting and packing of material matter were most widely discussed in approximately equal numbers of about 150. The number of sources handling three-dimensional problems (14) was extremely low. This is reflected in the fact that, in reality, three-dimensional problems are either standardized to fewer dimensions or cannot yet be solved by algorithmic solutions, as in cutting raw diamonds.

- About 80 per cent of the studies concentrated on either bin packing or, more dominantly, cutting stock types (over 50% of all cases). This may be due to the limitation of the area of investigation, but also arises from the numerous actual phenomena which are represented by this type[2].

- The frequency with which the literature addresses these general problem types does not allow for conclusions as to their actual frequency in reality. Furthermore, some sources refer to the same actual problem (so e.g. Gehring et al. 1979, Dyckhoff/Gehring 1982 and 1988). Nevertheless, the frequency might be used as a 'proxy' (index) for the importance, difficulty and variety of the respective type.

From the 308 analysed problems in the literature we therefore have 10 general types. In a more thorough examination of the grouped sources, it is possible to distinguish a series of further common properties, here named **additional characteristics** and presented in Table 5.2 under the following symbols:

⊡ The characteristic exists in all of the examined problem situations under the respective type.

⊟ The characteristic is not relevant for the respective type.

☐ The characteristic exists in few or none of the examined situations under the respective type.

1) In the case studied by Gehring et al. (1990) the original problem is a BP3-problem. It is, however, solved as a multiple KS3-problem.

2) Cf. Chapter 7.

additional characteristics				BP1	CS1	KS1	BP2	CS2	KS2	PL2	CS3	KS3	PL3
				general types									
quantity measurement			discrete	●		●	●	●	●	●	●	●	●
large objects	figure	form	rectangular	—	—	—	●		●	●	●		●
	assortment		homogeneous							●			●
		number/ figure	1						●	●			●
			limited (>1)			●						●	
		sequence	no order							●		●	●
small items	figure	form	rectangular	—	—	—					●	●	●
		orientation	90° turn permitted	—	—	—							●
	availability	number per figure	limited	●	●	●	●	●			●		●
		sequence	no order						●	●			●
		dates	same dates			●		●	●	●	●		●
pattern restrictions	geometric characteristics	no distance restrictions		●					●				●
		no frequency restrictions		●					●	●			●
		separation restrictions	orthogonal patterns	—	—	—					●	●	●
			nested patterns	—	—	—							●
	operational characteristics	number of stages	single-stage processes	●			●		●	●	●		●
		sequence	optional	●		●	●		●	●	●	●	●
		frequency	optional number of patterns	●		●	●		●	●	●	●	●
	dynamics of allocation	off-line problems (static)						●	●	●		●	●
		off- or on-line problems (static)		●		●							
objectives			value maximization										●
status of information			deterministic							●		●	●
variability			fixed dates	●		●			●	●	●	●	●
solution methods	object-orientated			●			●						
	object-orientated or single-patterned					●			●			●	
	single-patterned									●			●
	single- or multi-patterned				●			●			●		

Table 5.2: Overview of additional characteristics for general types in the analysed literature

One should not necessarily consider those additional properties found in the investigation as 'typical' for the respective C&P-type because the examined material is too limited for a definite assertion in this respect. There are, however, some rather important observations to be made in connection with Table 5.2:

- Six of the types, namely the one-dimensional bin packing type (BP1), the one- and two-dimensional knapsack types (KS1 and KS2), the two-dimensional pallet loading type (PL2) and all three-dimensional types (CS3, KS3 and PL3) possess a significant number of additional properties. This leads to the conclusion that in the sources these types are predominantly handled as so-called 'standard' or 'basic' types. A standard type is restricted - in respect to problem complexity - to the simplest of all possible properties (so-called standard properties), by which reality is idealized to an abstract, theoretical concept of well-defined mathematical structure, and for which solution procedures exist or can be developed. It is by the modification or extension of solution procedures for this sort of "simple", theoretical structure that solutions for more complex, practical problems can be developed. In this way standard types represent special types of an actual problem characterized by standard properties.

- Quantity measurement of objects and items is discrete in almost all types. An exception is the one-dimensional cutting stock type (CS1), in which both continuous and discrete quantity measurements appear (the former also being called '1.5-dimensional').

- Problems of irregularly-shaped large objects are treated in the examined literature in only two-dimensional cutting stock- and three-dimensional knapsack types (CS2 and KS3).

- Fully homogeneous assortments of large objects are only found in pallet loading types (PL2 and PL3). Since the item assortment in these types is also homogeneous, there is a representative solution of one object for all situations so that the number of objects per figure is one.

- In all of the knapsack types (KS1, KS2 and KS3) the number of objects per figure is limited. The 'classical' form, with only one object, is generally found in two-dimensional cases.

- All of the two-dimensional types also refer to situations with non-rectangular items. This property does not appear in three-dimensional types.

- Most of the pattern restrictions are found in one- and two-dimensional cutting stock types (CS1 and CS2). In fact, qualifications respective to pattern sequence and frequency occur only with these types.

- Dynamic problems are represented in one- and three-dimensional cutting stock types (CS1 and CS3) and in the two-dimensional bin packing type (BP2).

- Identical objectives for problems of the same type are exclusively found in the three-dimensional pallet loading type (PL3); this, however, is due to the fact that this example was analysed in only two of the sources.

- All contributions of pallet loading types (PL2 and PL3) deal with a deterministic status of information, whereas the other types (except KS3) also cater for a stochastic/random status of information.

- Bin packing types are solved with object-orientated methods, knapsack types with object- or single-pattern methods, pallet loading types with single-pattern methods, and cutting stock types with single- or multiple pattern methods.

5.2.2. Special Types

Special types of various degrees of abstraction are formed by further subdividing the above given general types. To do so, it is necessary to develop additional type-defining characteristics. Considering the great variety of actual C&P-problems, in principle this must be possible for every general type. But as discussed in the previous section, in the literature eight of the general types are handled with a great similarity in their characteristics (BP1, CS3, KS1, KS2, KS3, PL2 and PL3) or not at all (BP3). This means that there is no need for further subdividing, what does not mean that corresponding special types do not exist.[1] For example, it is possible to consider the seven general types mentioned above as special types, characterized by the standard properties of Table 5.2. Further investigation deals with the remaining three general types:

- two-dimensional bin packing type (BP2)
- one-dimensional cutting stock type (CS1)
- two-dimensional cutting stock type (CS2)

1) It should be reemphasized that this hierarchy of types is by no means complete. It is to be considered a theoretical framework which, if necessary, can/must be expanded upon and refined.

(1) Two-dimensional bin packing type (BP2)

The sources for BP2-types can be classified into five special types, partially of varying degrees of abstraction (see Figure 5.2).

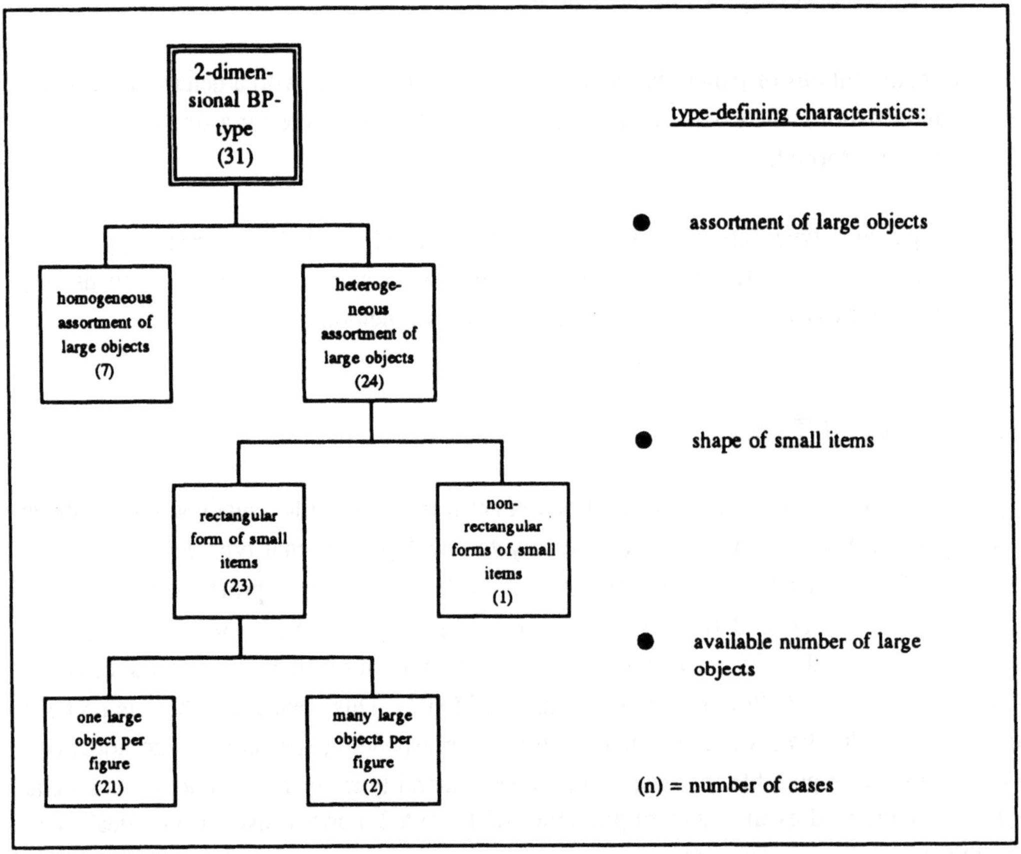

Figure 5.2: Special BP2-types

The results of the empirical study suggest considering the assortment of large objects as the first additional type-defining attribute. It is significant in that its varying properties distinctly influence the selection complexity with regard to large objects.

In the case of a homogeneous assortment the objects are only available in standard sizes of identical measurements, thus eliminating any alternatives for selection. In heterogeneous assortments the objects are available in various measurements, which leads to the problem of selecting those which should be employed for the C&P-process.

In the case of heterogeneous assortments the form of small items can serve as a further additional type-defining attribute. It significantly influences the complexity of the layout problem because the more irregular the form the more difficult the pattern design. For example, by now there is no computer program in the textile industry which provides a satisfactory layout for the many irregular pattern shapes for clothing.[1] Consequently, in heterogeneous assortments two further special types appear, namely the rectangular and the non-rectangular forms.

It is possible to further classify the special type of rectangular form by considering the available number of large objects. Depending on whether there is only one object per figure available or more, various implications appear for the selection of objects. The first case describes a special type in which one object of a particular width and infinite length must be assigned a predetermined number of rectangles of a certain length and width, so that the least possible 'length' is wasted. This problem, often under the name 'strip packing', is so idealized by assuming that instead of one object of infinite length, an infinite number of objects of all possible finite lengths (one per length) exists, from which the shortest is to be chosen for assignment. In this sort of structure there is a degree of choice concerning the assortment of large objects (assortment problem). The second case defines a special type with several large objects per figure of which the measurements are predetermined.

(2) One-dimensional cutting stock type (CS1)

Based on the results of empirical investigation the CS1-type can be divided into four special types (see Figure 5.3).

To begin with, the type of quantity measurement of large objects can serve as a type-defining attribute. Differences in its properties, discrete or continuous, do not necessarily influence the solution possibilities of a problem, because in actual decision-making, continuous quantity measurements are often assumed even for discrete problems and non-integer solutions are simply rounded.[2] They do, however, allow for conclusions concerning the problem context, clarifying specific characteristics which must be considered when developing a solution. One can generally conclude that those cases concerning continuous quantity measurement are problems of cutting rolls of material. In the case of discrete quantity measurements such assertions are not possible.

The type with discrete quantity measurement can be further differentiated with respect to the attribute 'assortment of large objects' (homogeneous or heterogeneous). Differences in its

1) Cf. Hummler (1989) p. 23.
2) Cf. Dyckhoff (1987) p. 97.

properties influence the complexity of selecting large objects in a manner similar to the BP2-type mentioned above.

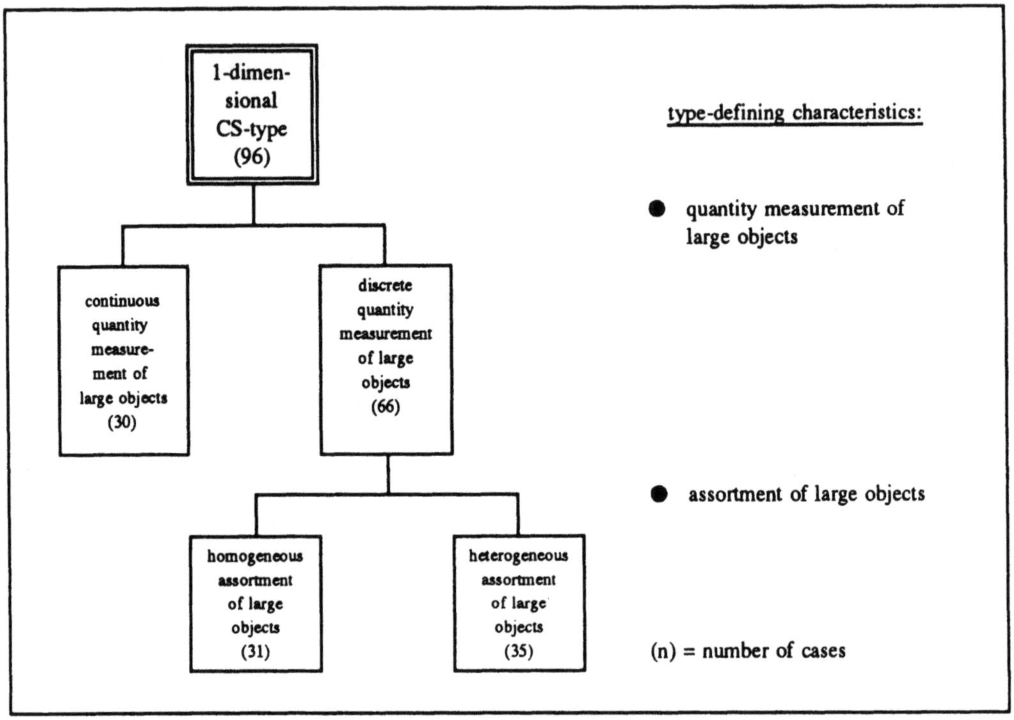

Figure 5.3: Special CS1-types

(3) Two-dimensional cutting stock type (CS2)

The investigation of sources on the CS2-type resulted in the classification of six special types (see Figure 5.4).

The first step is to differentiate between small items of rectangular or non-rectangular (in particular: irregular-shaped) forms. The significance of these differences on solutions for layout problems is referred to in the above paragraph on the BP2-type.

In CS2-types with rectangular small items the available number of large objects can serve as a type-defining attribute. Implications for object selection vary according to whether one or more large objects are available per figure.[1]

1) Refer to the comment on special BP2-types.

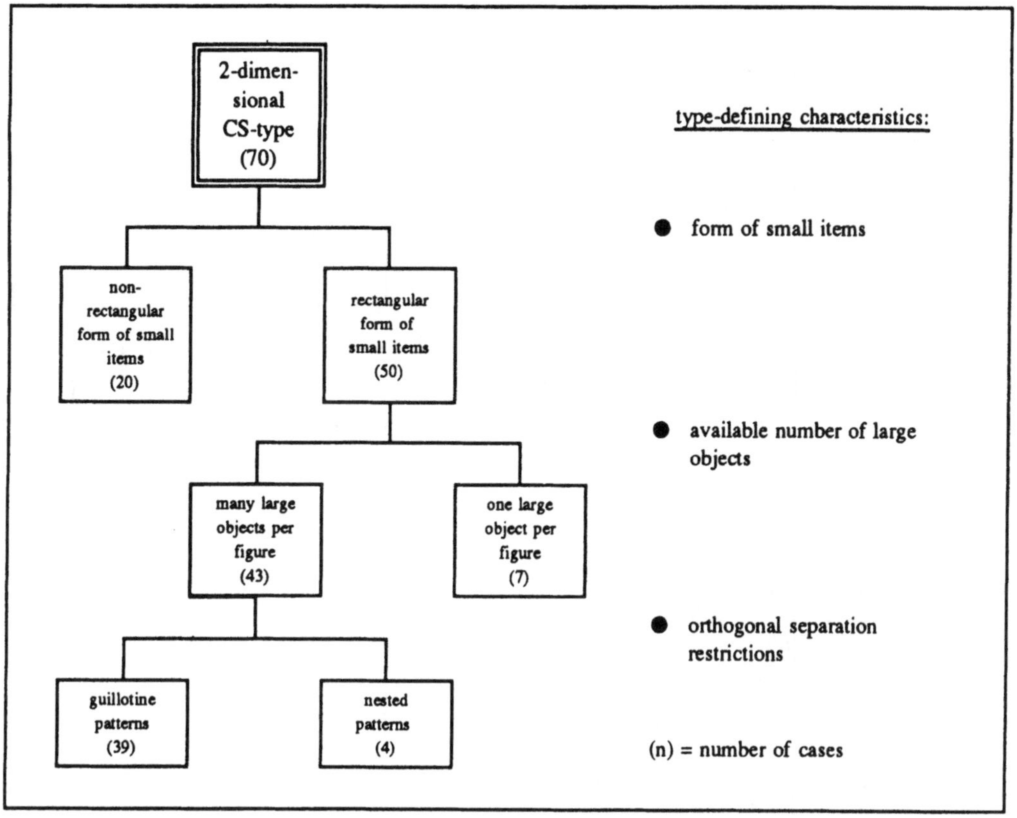

Figure 5.4: Special CS2-types

Special types in which there are several objects per figure can be further classified as to whether only guillotine patterns are allowed or nested patterns as well. The importance of this attribute results from its influence on the mathematical complexity of the layout problem. If only guillotine patterns are allowed in which the object is cut into parts by uninterrupted cuts, it will be possible to solve the problem by treating it as a series of interdependent, one-dimensional knapsack problems, which greatly simplifies the problem complexity.[1] In the case of nested patterns this simplification is not possible, and those problems are therefore comparatively more complex.

5.2.3. Summarized Description of the Hierarchy of Types

Figure 5.5 presents an overview of the deduced C&P problem types derived with an indication of the frequency with which they appear in the given literature. The general types are depicted by a double rim and special types by a single rim.

1) This procedure was developed by Gilmore/Gomory (1965). Two-dimensional C&P problems with guillotine patterns are in fact '1+1-dimensional'.

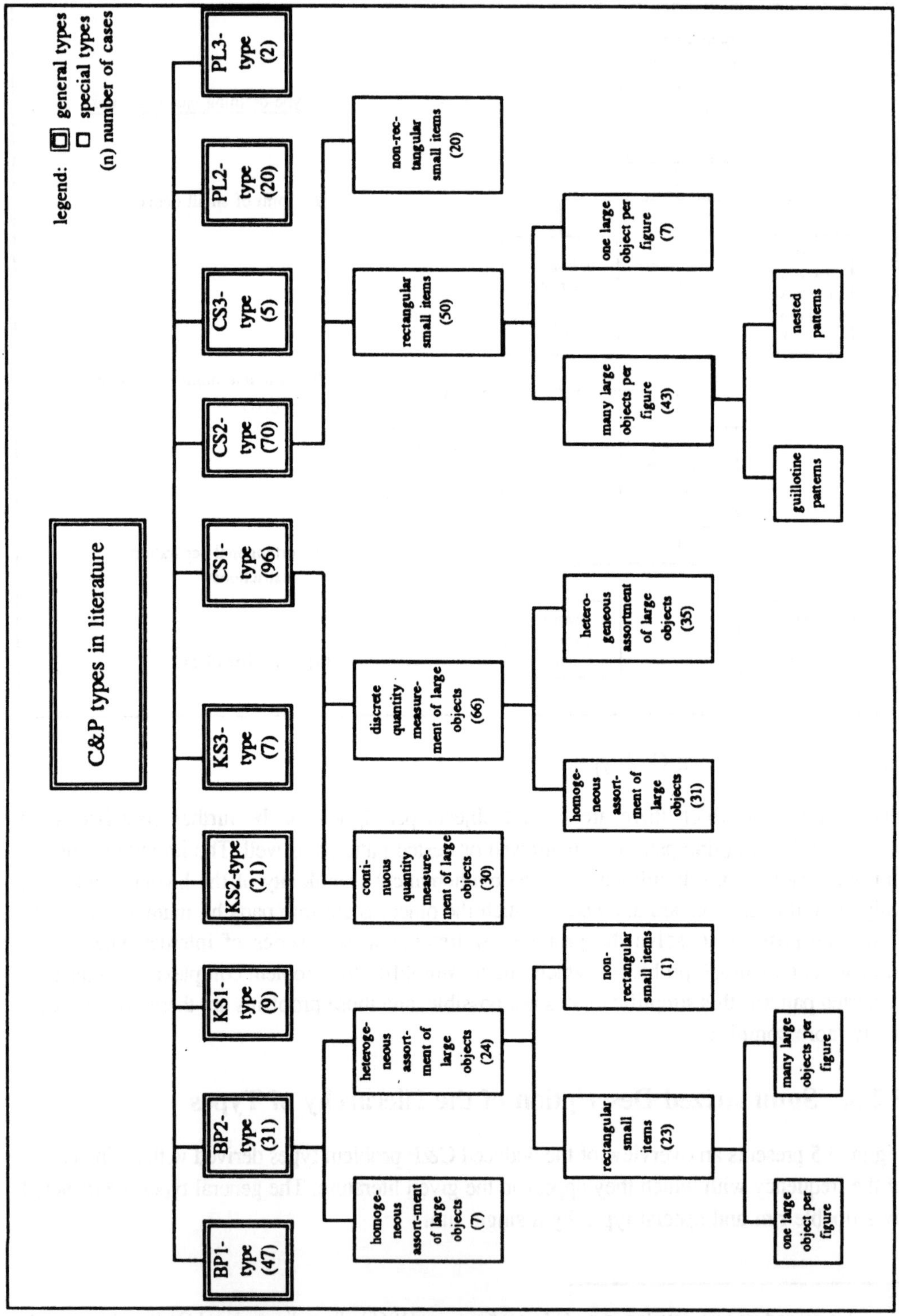

Figure 5.5: Hierarchy of types for C&P problems in the analysed literature

5.3. Properties of the Derived Problem Types

The following chapters will group the C&P problems found in the literature with respect to the developed hierarchy of types and compare the properties of those characteristic attributes not used in the type-definition. Two qualifications apply to the evaluation of these **type-describing properties**:

- In the examined sources some of the characteristics based on the logical structure could not be judged because of lack of sufficient information. This is true for the relation in size[1] of objects and items, the object orientation and the dates for the objects[2]. Some properties such as 'multi-dimensionality' do not occur in the sources analysed. For these reasons we slightly changed the catalogue presented in Chapter 4 with respect to the characteristics based on the logical structure. Table 5.3. describes this.

- Investigating properties of reality-based attributes proves to be meaningful only in application-orientated papers and case studies. Only in these sources sufficient exploitable information is available.

The comparison of the contributions is based on the developed hierarchy of types. The following procedure is as follows:

- First, those properties are presented which all sources belonging to the same specific type have in common. Included here is a recapitulation of the respective type-defining characteristics. The tables employ the structure of Table 5.3 and use the following symbols:

BOLD CAPITAL LETTERS (shaded box)	Characteristic is type-defining.
small letters (shaded box)	Characteristic is represented in all of the examined sources, but not as type-defining.
small letters	Characteristic is either not at all or only partially represented in the examined sources.
—	Characteristic is not relevant to the respective type.

1) Naturally the size is indirectly determined also by the attribute 'assortment'.

2) This does not mean, however, that these characteristics are not essential for C&P problem solutions. The relation in size of items and objects, for example, can play an important role in the solutions; cf. Goulimis (1990).

- Second, a survey is given of all the sources analyzed using overview matrices which assign to the individual sources (rows of the matrices) their properties (columns of the matrices). This structure thus allows a direct comparison of all contributions of a particular type. The contributions themselves are arranged according to the year of publication and within each year alphabetically by the authors' names. Sources containing more than one sort of problem appear on the list several times with numerical specifications. Information is also given as to whether the sources are theoretical/methodic, application-orientated, or case studies.The following symbols are used in the overview practices:

  [•] Property exists.

  [-] Property is not relevant.

  [ ] Property is non-existent or not evident in source.

- To clarify the overview, characteristics based on the logical structure and reality-based characteristics are examined separately.

attributes			properties			
dimensionality			1-dimensional	2-dimensional	3-dimensional	
type of assignment			Type II	Type III	Type IV	
large objects	type of quantity measurement		discrete	continuous		
	figure	form	rectangular	non-rectangular		
	assortment	homogeneous	fixed	variable		
		heterogeneous	fixed	variable		
	availability	number/figure	1	limited	unlimited	mixed
		sequence	no order	partial order	complete order	
small items	type of quantity measurement		discrete	continuous		
	figure	form	rectangular	non-rectangular		
		orientation	fixed	optional	90° turn (parallel)	90° turn (optional)
	assortment	homogeneous	fixed	variable		
		heterogeneous	fixed	variable	few objects per figure	many objects per figure
	availability	number/figure	limited	unlimited		
		sequence	no order	partial order	complete order	
		dates	identical time references	different time references		
pattern restrictions	geometric characteristics	distancing restrictions	none	between items	to the edges	
		frequency limitations	none	limited combinations of figures	limited number of figures	limited number of items
		separation restrictions	non-orthogonal patterns	nested patterns	guillotine patterns (limited stages)	guillotine patterns (unlimited stages)
	operational characteristics	number of stages	single-stage processes	multi-stage processes		
		pattern sequence	no restrictions	with restrictions		
		pattern frequency	no restrictions	limited number of different patterns	minimum number of identical patterns	
	dynamics of allocation		dynamic	static: off-line	static: on-line	
objectives			input minimization	trim loss minimization	value maximization	others
status of information			deterministic	stochastic		
variability of data			fixed data	variable data		
solution methods			object-orientated methods	single-patterned methods	multi-patterned methods	

Table 5.3: System of characteristics based on logical structure according to their portrayal in literature

6. Bin Packing Types (BP)

Most of the sources dealing with bin packing types are highly mathematical, primarily discussing the efficiency of algorithms. In this study, only those contributions were analysed which have a 'certain' reference to reality. However, there are border-line cases, which were assigned arbitrarily. More sources of bin packing types can be found in Appendix I, Part B under 'Mathematical Problems'.

6.1. One-dimensional Bin Packing Type (BP1)

(1) Common characteristics

Table 6.1 displays the common characteristics found in sources on the BP1-type. The following common characteristics can be identified aside from the type-defining ones:

- discrete quantity measurement of objects and items,
- fixed assortments of objects and items,
- limited number of small items per figure,
- no geometric characteristics,
- single-stage processes,
- no restrictions regarding pattern sequence or frequency,
- fixed data,
- object-orientated methods.

(2) Comparative contrast of literary sources

The analysed contributions of the BP1-type are contrasted in Table 6.2a (Part 1), Table 6.2b (Part 2) and Table 6.2c (Part 3). The following points become clear in their analysis:

- 96% of the contributions are theoretical/methodical studies; the rest are case studies. No application-orientated studies appear.
- In majority, the large object assortments (83%) are homogeneous, meaning that they have 'identical measurements' .
- In 84% of the cases, the available number of large objects is considered unlimited.
- In respect to the sequence of large objects, there are no restrictions in the most cases (96%).
- The small items, with only one exception, all have the same time reference.
- Only single objectives appear, whereby the property 'input minimization' dominates with 88%.
- 33% of the sources handle on-line problems, which are further distinguished by a complete order in the sequence of small items and a stochastic status of information.

attributes			properties			
dimensionality			1-DIMENSIONAL	2-dimensional	3-dimensional	
type of assignment			Type II	TYPE III	TYPE IV	
large objects	type of quantity measurement		discrete	continuous		
	figure	form	—	—		
	assortment	homogeneous	fixed	variable		
		heterogeneous	fixed	variable		
	availability	number/figure	1	limited	unlimited	mixed
		sequence	no order	partial order	complete order	
small items	type of quantity measurement		discrete	continuous		
	figure	form	—	—		
		orientation	—	—	—	—
	assortment	homogeneous	fixed	variable		
		heterogeneous	fixed	variable	FEW OBJECTS PER FIGURE	many objects per figure
	availability	number/figure	limited	unlimited		
		sequence	no order	partial order	complete order	
		dates	identical time references	different time references		
pattern restrictions	geometric characteristics	distancing restrictions	none	between items	to the edges	
		frequency limitations	none	limited combinations of figures	limited number of figures	limited number of items
		separation restrictions	—	—	—	—
	operational characteristics	number of stages	single-stage processes	multi-stage processes		
		pattern sequence	no restrictions	with restrictions		
		pattern frequency	no restrictions	limited number of different patterns	minimum number of identical patterns	
	dynamics of allocation		dynamic	static: off-line	static: on-line	
objectives			input minimization	trim loss minimization	value maximization	others
status of information			deterministic	stochastic		
variability of data			fixed data	variable data		
solution methods			object-orientated methods	single-patterned methods	multi-patterned methods	

Table 6.1: Common characteristics of the BP1-type

properties → / sources ↓	type of sources: theoretical/methodic	type of sources: case study	large objects – assortment: heterogeneous	large objects – assortment: homogeneous	large objects – availability – number/figure: limited	large objects – availability – number/figure: unlimited	large objects – availability – sequence: no order	large objects – availability – sequence: complete order	small items – availability – sequence: no order	small items – availability – sequence: partial order	small items – availability – sequence: complete order	small items – availability – dates: identical time reference	small items – availability – dates: different time reference	dynamics of allocation – static: on-line	dynamics of allocation – static: off-line	objectives: input minimization	objectives: trim loss minimization (absolute)	objectives: cost minimization	objectives: value maximization	status of information: deterministic	status of information: stochastic
author(s) and year of publication																					
Garey et al. (1973)-1	●			●		●	●		●			●			●	●				●	
Garey et al. (1973)-2	●			●		●	●		●			●			●	●				●	
Johnson et al. (1974)-1	●			●		●	●		●			●			●	●				●	
Johnson et al. (1974)-2	●			●		●	●		●			●			●	●				●	
Johnson (1974)-1	●			●		●	●		●			●			●	●				●	
Johnson (1974)-2	●			●		●	●		●			●			●	●				●	
Maruyama et al. (1977)	●		●		●		●		●			●			●	●					
Shapiro (1977)-1	●		●			●	●				●	●		●		●					●
Shapiro (1977)-2	●		●			●	●		●			●			●	●				●	
Garey et al. (1978)	●			●		●	●		●			●			●	●				●	

Aiello et al. (1980)	●			●	●		●		●			●			●	●				●	
Coffman et al. (1980b)	●			●		●	●				●	●		●		●					●
Fredericksen (1980)	●			●		●	●		●			●			●	●				●	
Liang (1980)	●			●		●	●				●	●		●		●					●
Yao (1980)	●			●		●	●				●	●		●				●			●
Baker/Coffman (1981)	●			●		●	●		●			●			●	●				●	
Vega/Lueker (1981)	●			●		●	●				●	●		●		●					●
Knödel (1981)	●			●	●		●		●			●			●	●				●	
Rocha et al. (1981a)		●	●		●		●		●			●			●		●			●	
Rocha et al. (1981b)		●	●		●		●		●			●			●		●			●	
Shearer (1981)	●			●	●		●				●	●		●		●					●

Table 6.2a: BP1-problems (Part 1)

properties / sources: author(s) and year of publication	type of sources: theoretical/methodic	type of sources: case study	large objects – assortment: heterogeneous	large objects – assortment: homogeneous	large objects – availability – number/figure: limited	large objects – availability – number/figure: unlimited	large objects – availability – sequence: no order	large objects – availability – sequence: complete order	small items – availability – sequence: no order	small items – availability – sequence: partial order	small items – availability – sequence: complete order	small items – availability – dates: identical time reference	small items – availability – dates: different time reference	dynamics of allocation – static: on-line	dynamics of allocation – static: off-line	objectives: input minimization	objectives: trim loss minimization (absolute)	objectives: cost minimization	objectives: value maximization	status of information: deterministic	status of information: stochastic
Hoffmann (1982)	●			●		●	●				●	●		●		●					●
Karmarkar (1982)	●			●		●	●				●	●		●		●					●
Bentley et al. (1983)	●			●		●	●				●	●			●	●				●	
Coffman et al. (1983)	●			●		●	●				●		●	●		●					●
Knödel (1983)	●			●		●	●				●	●		●		●					●
Rayward-Smith/Shing (1983)	●			●		●	●				●	●		●		●					●
Eilon (1984)	●			●		●	●				●	●			●	●				●	
Loulou (1984)	●			●		●	●				●	●			●	●				●	
Ong et al. (1984)-1	●			●		●	●		●			●			●				●	●	
Ong et al. (1984)-2	●			●		●	●				●	●			●				●	●	

Shor (1984)	●			●		●	●				●	●		●		●					●
Johnson/Garey (1985)	●			●		●	●				●	●			●	●				●	
Martel (1985)	●			●		●	●			●		●			●	●				●	
Rhee (1985)	●			●		●	●				●	●		●		●					●
Csirik et al. (1986)	●			●		●	●				●	●			●	●				●	
Csirik/Galambos (1986)	●			●		●	●		●			●			●	●				●	
Csirik/Mate (1986)	●			●		●	●				●	●			●	●				●	
Csirik (1986)	●			●		●	●				●	●		●		●					●
Friesen/Langston (1986)	●		●			●		●			●	●			●			●		●	
Galambos (1986)	●			●		●	●				●	●			●	●				●	
Shor (1986)	●			●		●	●				●	●		●		●					●

Table 6.2b: BP1-problems (Part 2)

properties / sources	type of sources		large objects						small items					dynamics of allocation		objectives				status of information	
			assortment		availability				availability					static							
					number/figure		sequence		sequence			dates									
author(s) and year of publication	theoretical/methodic	case study	heterogeneous	homogeneous	limited	unlimited	no order	complete order	no order	partial order	complete order	identical time reference	different time reference	on-line	off-line	input minimization	trim loss minimization (absolute)	cost minimization	value maximization	deterministic	stochastic
Murgolo (1987)	●		●			●	●		●			●			●	●				●	
Hall (1988)-1	●			●		●	●		●			●			●	●				●	
Hall (1988)-2	●			●		●	●				●	●			●	●				●	
Csirik (1989)	●		●		●			●	●			●			●	●				●	
Csirik/Imreh (1989)	●			●	●		●		●			●			●	●				●	

Table 6.2c: BP1-problems (Part 3)

6.2. Two-dimensional Bin Packing Types (BP2)

6.2.1. BP2-Type with a Heterogeneous Assortment of Large Objects

The BP2-types with a heterogeneous assortment of large objects can be divided into three groups in respect to the developed hierarchy of types. The first group, which represents 88% of the contributions, consists of rectangular small items and only one large object per figure. The second group, which consists of rectangular small items and several objects per figure appears only twice and the third with irregularly-shaped forms only once. In order to simplify the description, all three groups are analysed together.

(1) Common characteristics

The common characteristics of the BP2-type with a heterogeneous assortment of large objects are identified in Table 6.3. The heterogeneous assortments of the objects are, without exception, fixed. So, this property is marked as type-defining characteristic.

The following common properties assist the type-defining ones:

- discrete quantity measurement of objects and items,
- rectangular large objects,
- fixed assortments of items,
- limited number of small items per figure,
- no distancing restrictions or frequency limitations,
- single-stage processes,
- no restrictions in respect to pattern sequence or frequency,
- object-orientated methods.

In those cases where only one object per figure exists, two more correlations appear, namely:

- rectangular small items,
- identical time references.

(2) Comparative contrast of literary sources

The overview matrix of BP2-types with a heterogeneous assortment of objects (Table 6.4) is divided into three groups in relation to the hierarchy of types. A bold line separates sources of the various types.

Regarding those sources handling types with one object per figure the matrix shows that aside from the type-forming characteristics extensive correlations appear in relation to these properties:

- theoretical/methodical studies (100%),
- no given order of large objects (95%),
- no given order of small items (75%),
- nested patterns (86%),
- off-line problems (86%),
- input minimization (81%),
- deterministic status of information (86%),
- fixed data (91%).

attributes			properties			
dimensionality			1-dimensional	2-DIMENSIONAL	3-dimensional	
type of assignment			Type II	TYPE III	TYPE IV	
large objects	type of quantity measurement		discrete	continuous		
	figure	form	rectangular	non-rectangular		
	assortment	homogeneous	fixed	variable		
		heterogeneous	FIXED	variable		
	availability	number/figure	1	limited	unlimited	mixed
		sequence	no order	partial order	complete order	
small items	type of quantity measurement		discrete	continuous		
	figure	form	rectangular	non-rectangular		
		orientation	fixed	optional	90° turn (parallel)	90° turn (optional)
	assortment	homogeneous	fixed	variable		
		heterogeneous	fixed	variable	FEW OBJECTS PER FIGURE	many objects per figure
	availability	number/figure	limited	unlimited		
		sequence	no order	partial order	complete order	
		dates	identical time references	different time references		
pattern restrictions	geometric characteristics	distancing restrictions	none	between items	to the edges	
		frequency limitations	none	limited combinations of figures	limited number of figures	limited number of items
		separation restrictions	non-orthogonal patterns	nested patterns	guillotine patterns (limited stages)	guillotine patterns (unlimited stages)
	operational characteristics	number of stages	single-stage processes	multi-stage processes		
		pattern sequence	no restrictions	with restrictions		
		pattern frequency	no restrictions	limited number of different patterns	minimum number of identical patterns	
	dynamics of allocation		dynamic	static: off-line	static: on-line	
objectives			input minimization	trim loss minimization	value maximization	others
status of information			deterministic	stochastic		
variability of data			fixed data	variable data		
solution methods			object-orientated methods	single-patterned methods	multi-patterned methods	

Table 6.3: Common characteristics of the BP2-type with a heterogeneous assortment of large objects

sources / properties: author(s) and year of publication	type of sources: theoretical/methodic	type of sources: case study	large objects: availability: number/figure: limited: one large object	large objects: availability: number/figure: unlimited	large objects: availability: sequence: no order	large objects: availability: sequence: order	small items: shape: rectangular	small items: shape: non-rectangular	small items: shape: orientation: fixed	small items: shape: orientation: 90 degree-turns (parallel)	small items: availability: sequence: no order	small items: availability: sequence: order	small items: availability: dates: identical time reference	small items: availability: dates: different time reference	pattern restrictions: geometry: separation: non-orthogonal	pattern restrictions: geometry: separation: orthogonal: nested patterns	pattern restrictions: geometry: separation: orthogonal: guillotine patterns with unlimited stages	pattern restrictions: dynamics of allocation: static: on-line	pattern restrictions: dynamics of allocation: static: off-line	pattern restrictions: dynamics of allocation: dynamic (without reallocation)	objectives: input minimization	objectives: trim loss minimization (absolute)	status of information: deterministic	status of information: stochastic	variability: fixed data	variability: variable data
Baker et al. (1980)	●		●		●		●		●		●		●			●			●		●		●		●	
Brown (1980)	●		●		●		●		●		●		●			●			●		●		●		●	
Coffman et al. (1980a)	●		●		●		●		●		●		●			●			●		●		●		●	
Frederickson (1980)	●		●		●		●		●		●		●			●			●		●		●		●	
Hofri (1980)-1	●		●		●		●		●		●		●			●			●			●	●		●	
Hofri (1980)-2	●		●		●		●			●	●		●			●			●			●	●		●	
Hofri (1980)-3	●		●		●		●		●		●		●			●			●			●	●		●	
Sleator (1980)	●		●		●		●		●		●		●			●			●		●		●		●	
Baker et al. (1981)	●		●		●		●		●		●		●			●			●		●		●		●	
Golan (1981)-1	●		●		●		●		●		●		●			●			●		●		●		●	
Golan (1981)-2	●		●		●		●		●		●		●			●			●		●		●		●	
Bengtsson (1982)-1	●		●		●		●			●	●		●			●			●		●		●		●	
Brown (1982)	●		●		●		●		●			●	●			●		●			●			●	●	

Baker/Schwarz (1983)	●		●		●		●		●		●		●			●			●		●		●		●	
Baker et al. (1983)-2	●		●		●		●			●	●		●			●			●		●		●		●	
Chazelle (1983)	●		●		●		●		●		●		●			●			●		●		●		●	
Rayward-Smith/Shing (1983)-2	●		●		●		●		●		●		●			●			●		●		●		●	
Rode/Rosenberg (1987)	●		●		●		●		●		●		●				●		●		●		●		●	
Bartholdi et al. (1989)	●		●		●		●		●			●	●					●			●					●
Coffman/Lagarias (1989)	●		●		●		●				●		●			●			●		●		●		●	
Coffman/Shor (1990)-2	●		●			●	●		●			●	●				●			●		●		●		
Larsen/Mikkelsen (1980)		●		●		●	●					●		●						●		●	●		●	
Bengtsson (1982)-2	●			●	●		●			●	●		●			●			●		●		●		●	
Qu/Sanders (1987)	●			●	●			●			●		●		●				●			●	●		●	

Table 6.4: BP2-problems with a heterogeneous assortment of large objects

6.2.2. BP2-Type with a Homogeneous Assortment of Large Objects

(1) Common characteristics

Table 6.5 displays the common characteristics of the BP2-type with a homogeneous assortment of large objects. In all cases the heterogeneous assortment of the objects is fixed. Therefore, this property is marked as type-defining characteristic.

The following similarities can be found aside from the type-defining characteristics:

- discrete quantity measurement of objects and items,
- rectangular objects and items,
- fixed assortments of items,
- limited number of small items per figure,
- no distancing restrictions or frequency limitations,
- nested patterns,
- single-stage processes,
- no restrictions concerning pattern sequence or frequency,
- object-orientated methods.

(2) Comparative contrast of literary sources

In consideration of the overview matrix of the BP2-type with a homogeneous assortment of large objects (Table 6.6), these properties appear most frequently:

- theoretical/methodical studies (86%),
- no given order of large objects (86%),
- small items of identical time reference (86%),
- off-line problems (71%),
- input minimization (71%),
- deterministic status of information (85%),
- fixed data (85%).

Interestingly, it's in the only case study (Bookbinder/Higginson 1986) where are containing the most deviations from these 'standard properties'.

attributes			properties			
dimensionality			1-dimensional	2-DIMENSIONAL	3-dimensional	
type of assignment			Type II	TYPE III	TYPE IV	
large objects	type of quantity measurement		discrete	continuous		
	figure	form	rectangular	non-rectangular		
	assortment	homogeneous	FIXED	variable		
		heterogeneous	fixed	variable		
	availability	number/figure	1	limited	unlimited	mixed
		sequence	no order	partial order	complete order	
small items	type of quantity measurement		discrete	continuous		
	figure	form	rectangular	non-rectangular		
		orientation	fixed	optional	90° turn (parallel)	90° turn (optional)
	assortment	homogeneous	fixed	variable		
		heterogeneous	fixed	variable	FEW OBJECTS PER FIGURE	many objects per figure
	availability	number/figure	limited	unlimited		
		sequence	no order	partial order	complete order	
		dates	identical time references	different time references		
pattern restrictions	geometric characteristics	distancing restrictions	none	between items	to the edges	
		frequency limitations	none	limited combinations of figures	limited number of figures	limited number of items
		separation restrictions	non-orthogonal patterns	nested patterns	guillotine patterns (limited stages)	guillotine patterns (unlimited stages)
	operational characteristics	number of stages	single-stage processes	multi-stage processes		
		pattern sequence	no restrictions	with restrictions		
		pattern frequency	no restrictions	limited number of different patterns	minimum number of identical patterns	
	dynamics of allocation		dynamic	static: off-line	static: on-line	
objectives			input minimization	trim loss minimization	value maximization	others
status of information			deterministic	stochastic		
variability of data			fixed data	variable data		
solution methods			object-orientated methods	single-patterned methods	multi-patterned methods	

Table 6.5: Common characteristics of the BP2-type with a homogeneous assortment of large objects

properties / sources	type of sources		large objects		small items							pattern restrictions					objectives			status of information		variability	
			availability		shape		availability					geometry		dynamics of allocation									
			sequence		orientation		sequence			dates			frequency	static									
author(s) and year of publication	theoretical/methodic	case study	no order	partial order	fixed	90 degree-turns (parallel)	no order	partial order	complete order	identical time reference	different time reference	distance to the edges	limited combinations of figures	on-line	off-line	dynamic (with reallocation)	input minimization	trim loss minimization (absolute)	cost minimization	deterministic	stochastic	fixed data	variable data
Chung et al. (1982b)	●		●		●		●			●					●		●			●		●	
Israni/Sanders (1985)	●		●			●	●			●					●			●		●		●	
Bookbinder/Higginson (1986)		●	●					●			●	●	●			●			●	●			●
Berkey/Wang (1987)	●		●		●		●			●					●		●			●		●	
Frenk/Galambos (1987)	●		●		●		●			●					●		●			●		●	
Coppersmith/Raghavan (1989)	●			●					●	●			●	●			●				●	●	
Coffman/Shor (1990)-1	●		●				●			●					●		●			●		●	

Table 6.6: BP2-problems with a homogeneous assortment of large objects

6.3. Actual Bin Packing Problems

Table 6.7 presents the properties of reality-based attributes appearing in application-orientated studies or case studies for bin packing types. There are three cutting problems and one packing problem. In solutions for the cutting problems, connections with inventory planning or scheduling are considered; for the packing problem balance conditions for the object to be packed (the aeroplane) are of importance in planning. Software for practical application exists for three of the four problems described.

	properties		kind of objects and items		planning situation														software	
					cutting problems											packing problems				
	sources / author(s) and year of publication	branch of industry	large objects	small items	isolated cutting planning	with order management	with timing	with capacity planning	with sequencing	with choice of manufacturing methods	with numerical control	with inventory planning	with material requirement planning	with procurement planning	others	isolated packing planning	with packaging planning	others	practical application	simulation
1-dimensional	Rocha et al. (1981a)		telephone cable	telephone cable								●							●	
1-dimensional	Rocha et al. (1981b)		telephone cable	telephone cable								●							●	
2-dimensional	Larsen/Mikkelsen (1980)	cargo aircraft	cargo compartments	containers and pallets														with balancing	●	
2-dimensional	Bookbinder/Higginson (1986)	paper industry	corrugated cardboard	corrugated cardboard					●											●

Table 6.7: Actual bin packing problems

7. Cutting Stock Types (CS)

7.1. One-dimensional Cutting Stock Types (CS1)

7.1.1. CS1-Type with Continuous Quantity Measurement of Large Objects

(1) Common characteristics

Only a few common characteristics arise in the sources aside from the type-defining ones found in Table 7.1, namely:

- fixed figure of objects and items,
- continuous quantity measurement of small items,
- limited number of items per figure,
- no given sequence of large objects,
- pattern-orientated methods.

(2) Comparative contrast of literary sources

Sources for the CS1-type with continuous quantity measurement are presented in the following tables:

- Table 7.2a: Without pattern restrictions, Part 1
- Table 7.2b: Without pattern restrictions, Part 2
- Table 7.2c: Only pattern restrictions, Part 1
- Table 7.2d: Only pattern restrictions, Part 2

The following ascertainments were made from the investigation:

- A significant number (87%) of the contributions are either application-orientated studies or case studies.

- In 73% of the studies the large objects have different measurements (i.e. a heterogeneous assortment).

- More than a half (60%) of the sources present an unlimited number of available objects per figure.

- There are no restrictions in reference to the sequence of small items in 80% of the cases.

- All of the small items have identical time references, except for one case, which is the only dynamic problem. The remaining are static off-line problems.

- Distancing restrictions and frequency limitations appear in only a few (23%) studies. Separation restrictions are of no significance in one-dimensional problems and therefore do not appear.

- Concerning operational characteristics, problems with optional pattern sequence (93%) and frequency (90%) are dominant.

- Both single-stage and multi-stage processes appear, although the former with 67% is more common.

- 93% of the studies include objectives orientated on trim loss or use of large objects such as input, trim loss (relative or absolute) or cost minimization. The most cases (77%) are also accompanied by further objectives, often change-over cost or inventory cost minimization.

- A deterministic status of information exists in most of the problems.

- In 60% of the sources the data are variable to a certain extent, that is, over- or underfulfilling in the orders on hand for small items may occur.

- A significant majority of the problems (93%) are at least partly solved by multi-patterned procedures.

attributes			properties			
dimensionality			1-DIMENSIONAL	2-dimensional	3-dimensional	
type of assignment			Type II	TYPE III	TYPE IV	
large objects	type of quantity measurement		discrete	CONTINUOUS		
	figure	form	—	—		
	assortment	homogeneous	fixed	variable		
		heterogeneous	fixed	variable		
	availability	number/figure	1	limited	unlimited	mixed
		sequence	no order	partial order	complete order	
small items	type of quantity measurement		discrete	continuous		
	figure	form	—	—		
		orientation	—	—	—	—
	assortment	homogeneous	fixed	variable		
		heterogeneous	fixed	variable	few objects per figure	MANY OBJECTS PER FIGURE
	availability	number/figure	limited	unlimited		
		sequence	no order	partial order	complete order	
		dates	identical time references	different time references		
pattern restrictions	geometric characteristics	distancing restrictions	none	between items	to the edges	
		frequency limitations	none	limited combinations of figures	limited number of figures	limited number of items
		separation restrictions	—	—	—	—
	operational characteristics	number of stages	single-stage processes	multi-stage processes		
		pattern sequence	no restrictions	with restrictions		
		pattern frequency	no restrictions	limited number of different patterns	minimum number of identical patterns	
	dynamics of allocation		dynamic	static: off-line	static: on-line	
objectives			input minimization	trim loss minimization	value maximization	others
status of information			deterministic	stochastic		
variability of data			fixed data	variable data		
solution methods			object-orientated methods	single-patterned methods	multi-patterned methods	

Table 7.1: Common characteristics of the CS1-type with continuous quantity measurement

sources / properties		type of sources			large objects					small items				objectives							status of information		variability		solution approaches	
					assortment		availability			availability															pattern-orientated	
							number/figure			sequence		dates														
sources	author(s) and year of publication	theoretical/methodic	application-orientated	case study	heterogeneous	homogeneous	limited	unlimited	mixed	no order	partial order	identical time reference	different time reference	input minimization	trim loss minimization (absolute)	trim loss minimization (relative)	cost minimization	cutting cost minimization	inventory cost minimization	others	deterministic	stochastic	fixed data	variable data	single-patterned	multi-patterned
	Vajda (1958)		●		●			●		●		●		●							●		●			●
	Förstner (1959)-1			●	●			●		●		●			●						●		●			●
	Förstner (1959)-2			●	●			●		●			●			●			●		●		●			●
	Förstner (1963)		●		●			●		●		●					●		●		●		●			●
	Meerendonk et al. (1963)		●		●			●		●		●		●							●		●			●
	Gilmore (1966)	●			●			●		●		●					●				●		●			●
	Poirier (1967)			●	●			●			●	●				●					●			●		●
	Caruso/Kokat (1973)			●	●			●			●		●			●								●		●
	Goswami (1973)			●	●				●	●		●				●					●		●			●
	Beged-Dov (1974)		●			●	●			●		●				●					●			●		●
	Hardley (1976)			●	●			●		●		●		●				●			●			●		●

Haessler (1978)-1			●	●		●			●		●					●	●			●			●		●
Haessler (1978)-2			●		●		●		●		●					●	●			●			●		●
Schepens (1978)			●	●			●		●		●					●	●			●			●		●
Gilmore (1979)	●			●			●		●		●					●				●		●		●	●
Haessler (1979)			●		●		●		●		●					●	●			●			●		●
Dyckhoff (1981)		●		●				●	●		●					●				●		●			●
Dyckhoff/Gehring (1982)		●		●				●	●		●								● 1)	●			●		●
Johnston (1982)		●			●		●			●	●			●						●			●		●
Vonderembse/Haessler (1982)		●		●		●			●		●					●				●			●		●
Haessler/Talbot (1983)		●		●			●			●		●		●			●			●			●		●
Haessler (1983)		●			●		●		●		●			●			●			●			●		●
Bartmann (1986)			●	●				●		●	●			●				●	● 2)		●		●	●	
Rijckaert (1986)	●			●				●	●		●			●						●		●		●	●
Tabucanon/ Lertcharoensombat (1986)		●			●		●		●		●					●				●		●		●	

1) net return maximization
stock value maximization
2) volume throughput maximization

Table 7.2a: CS1-problems with continuous quantity measurement of large objects (without pattern restrictions, Part 1)

properties / sources: author(s) and year of publication	type of sources: theoretical/methodic	type of sources: application-orientated	type of sources: case study	large objects: assortment: heterogeneous	large objects: assortment: homogeneous	large objects: availability: number/figure: limited	large objects: availability: number/figure: unlimited	large objects: availability: number/figure: mixed	small items: availability: sequence: no order	small items: availability: sequence: partial order	small items: availability: dates: identical time reference	small items: availability: dates: different time reference	objectives: input minimization	objectives: trim loss minimization (absolute)	objectives: trim loss minimization (relative)	objectives: cost minimization	objectives: cutting cost minimization	objectives: inventory cost minimization	objectives: others	status of information: deterministic	status of information: stochastic	variability: fixed data	variability: variable data	solution approaches: pattern-orientated: single-patterned	solution approaches: pattern-orientated: multi-patterned
Dyckhoff/Gehring (1988)		●		●				●	●		●								● 1)	●			●		●
Nickels (1988)		●		●		●				●	●			●			●			●			●		●
Stadtler (1988)-2	●				●		●		●		●					●				●		●			●
Goulimis (1990)		●			●	●			●		●				●		●			●			●		●
Wäscher (1990)		●		●				●	●		●					●		●		●			●		●

1) net return maximization
stock value maximization

Table 7.2b: CS1-problems with continuous quantity measurement of large objects (without pattern restrictions, Part 2)

sources / properties	pattern restrictions																		
	geometry									organization						dynamics of allocation			
	distance		frequency			separation				number of stages		sequence		frequency of patterns		static		dynamic	
							orthogonal												
								guillotine patterns											
author(s) and year of publication	between items	to the edges	limited combinations of figures	limited number of figures	limited number of items	non-orthogonal	nested patterns	limited stages	unlimited stages	single-stage	multi-stage	unrestricted	restricted	limited number of different pattern types	lowest number of same pattern type	on-line	off-line	with reallocation	without reallocation
Vajda (1958)										●		●					●		
Förstner (1959)-1										●		●					●		
Förstner (1959)-2										●		●					●		
Förstner (1963)										●		●					●		
Meerendonk et al. (1963)					●					●		●					●		
Gilmore (1966)										●		●					●		
Poirier (1967)											●	●					●		
Caruso/Kokat (1973)										●		●						●	
Goswami (1973)		●									●	●					●		

Table 7.2c: CS1-problems with continuous quantity measurement of large objects (only pattern restrictions, Part 1)

properties	pattern restrictions														dynamics of allocation			
	geometry									organization								
	distance		frequency			separation				number of stages		sequence		frequency of patterns		static		dynamic	
							orthogonal											
								guillotine patterns										
sources / author(s) and year of publication	between items	to the edges	limited combinations of figures	limited number of figures	limited number of items	non-orthogonal	nested patterns	limited stages	unlimited stages	single-stage	multi-stage	unrestricted	restricted	limited number of different pattern types	lowest number of same pattern type	on-line	off-line	with reallocation	without reallocation
Beged-Dov (1974)										●		●					●		
Hardley (1976)										●			●	●			●		
Haessler (1978)-1										●		●		●			●		
Haessler (1978)-2		●								●		●		●			●		
Schepens (1978)											●	●					●		
Gilmore (1979)										●		●					●		
Haessler (1979)-2											●	●					●		
Dyckhoff (1981)											●	●					●		
Dyckhoff/Gehring (1982)										●		●					●		

Johnston (1982)			●		●						●	●					●		
Vonderembse/Haessler (1982)										●		●					●		
Haessler/Talbot (1983)		●	●								●		●				●		
Haessler (1983)										●		●					●		
Bartmann (1986)		●	●								●	●					●		
Rijckaert (1986)										●		●					●		
Tabucanon/ Lertcharoensombat (1986)										●		●					●		
Dyckhoff/Gehring (1988)										●		●					●		
Nickels (1988)											●	●					●		
Stadtler (1988)-2										●		●					●		
Goulimis (1990)					●						●	●					●		
Wäscher (1990)										●		●					●		

Table 7.2d: CS1-problems with continuous quantity measurement of large objects (only pattern restrictions, Part 2)

7.1.2. CS1-Types with Discrete Quantity Measurement of Large Objects

7.1.2.1. Discrete CS1-Type with a Homogeneous Assortment of Large Objects

(1) Common characteristics

Table 7.3 displays the common characteristics found in CS1-types with discrete quantity measurement and a homogeneous assortment of large objects. Because of the fact, that in all cases the homogeneous assortment of the objects is fixed, this property is marked as type-defining characteristic.

The following similarities existing aside the type-defining characteristics:

- no given order of large objects,
- discrete quantity measurement of small items,
- fixed figures of items,
- limited number of small items per figure,
- small items of identical time reference,
- no restrictions concerning pattern frequency,
- (static) off-line problems,
- pattern-orientated methods.

(2) Comparative contrast of literary sources

The analysed contributions of the above specified CS1-type are contrasted in Table 7.4a (Part1) and Table 7.4b (Part2), in which the following points become clear:

- About half (48%) of the studies are theoretical/methodical, 36% application-orientated and 16% case studies.

- The number of available large objects is primarily (94%) unlimited.

- In respect to the sequence of small items there are in most cases (84%) no restrictions. Those which do have restrictions specify either a partial (Karmarkar/Karp 1982) or a complete order.

- Distancing restrictions or frequency limitations occur in only 29% of the studies, concerning the distance to the edge or the number of small items. Separation restrictions are of no importance in one-dimensional problems.

- 90% of the problems handle single-stage processes.

- There is often (65%) only one objective in the process, whether it be input minimization, relative or absolute trim loss minimization or materials or space cost minimization. In certain cases installation costs also come into question.

- Exept in one case, the problems have a deterministic status of information.

- Fixed data exist in 70% of the sources.

- 84% of the solution approaches are multi-patterned.

attributes			properties			
dimensionality			1-DIMENSIONAL	2-dimensional	3-dimensional	
type of assignment			Type II	TYPE III	TYPE IV	
large objects	type of quantity measurement		DISCRETE	continuous		
	figure	form	—	—		
	assortment	homogeneous	FIXED	variable		
		heterogeneous	fixed	variable		
	availability	number/figure	1	limited	unlimited	mixed
		sequence	no order	partial order	complete order	
small items	type of quantity measurement		discrete	continuous		
	figure	form	—	—		
		orientation	—	—	—	—
	assortment	homogeneous	fixed	variable		
		heterogeneous	fixed	variable	few objects per figure	MANY OBJECTS PER FIGURE
	availability	number/figure	limited	unlimited		
		sequence	no order	partial order	complete order	
		dates	identical time references	different time references		
pattern restrictions	geometric characteristics	distancing restrictions	none	between items	to the edges	
		frequency limitations	none	limited combinations of figures	limited number of figures	limited number of items
		separation restrictions	—	—	—	—
	operational characteristics	number of stages	single-stage processes	multi-stage processes		
		pattern sequence	no restrictions	with restrictions		
		pattern frequency	no restrictions	limited number of different patterns	minimum number of identical patterns	
	dynamics of allocation		dynamic	static: off-line	static: on-line	
objectives			input minimization	trim loss minimization	value maximization	others
status of information			deterministic	stochastic		
variability of data			fixed data	variable data		
solution methods			object-orientated methods	single-patterned methods	multi-patterned methods	

Table 7.3: Common characteristics of the CS1-type with discrete quantity measurement and a homogeneous assortment of large objects

properties	type of sources			large objects		small items		pattern restrictions							objectives						status of information		variability		solution approaches	
				availability		availability		geometry			organization														pattern-orientated	
				number/figure		sequence		distance		fre-quency	number of stages		sequence													
sources: author(s) and year of publication	theoretical/methodic	application-orientated	case study	limited	unlimited	no order	order	between items	to the edges	limited number of items	single-stage	multi-stage	unrestricted	restricted	input minimization	trim loss minimization (absolute)	trim loss minimization (relative)	cost minimization	cutting cost minimization	others	deterministic	stochastic	fixed data	variable data	single-patterned	multi-patterned
Metzger (1958)		●			●	●			●	●	●		●			●						●	●			●
Eilon (1960)			●		●	●					●		●			●			●		●		●			●
Kantorovich (1960)	●			●		●					●		●			●					●		●		●	
Kreko (1965)		●			●	●					●		●			●					●		●			●
Johns (1967)		●			●	●					●		●				●				●			●		●
Eilon/Christofides (1971)	●				●	●					●		●		●						●		●		●	
Haessler (1971)			●		●	●			●			●	●					●			●			●		●
Zimmermann (1971)	●				●	●			●		●		●			●					●		●			●
Eilon/Christofides (1972)	●				●	●					●		●		●						●		●		●	
Lev (1972)	●				●		●				●		●		●						●		●		●	
Müller-Merbach (1973)	●				●	●					●		●			●					●		●			●
Coverdale/Wharton (1976)		●			●	●			●	●		●	●			●			●		●			●		●
Ahluwalia/Saxena (1978)		●		●		●					●		●							● 1)	●		●		●	●

1) total cost minimization

Table 7.4a: CS1-problems with discrete quantity measurement and a homogeneous assortment of large objects (Part 1)

properties / sources: author(s) and year of publication	type of sources: theoretical/methodical	type of sources: application-oriented	type of sources: case study	large objects – availability – number/figure: limited	large objects – availability – number/figure: unlimited	small items – availability – sequence: no order	small items – availability – sequence: order	pattern restrictions – geometry – distance: between items	pattern restrictions – geometry – distance: to the edges	pattern restrictions – geometry – frequency: limited number of items	pattern restrictions – organization – number of stages: single-stage	pattern restrictions – organization – number of stages: multi-stage	pattern restrictions – organization – sequence: unrestricted	pattern restrictions – organization – sequence: restricted	objectives: input minimization	objectives: trim loss minimization (absolute)	objectives: trim loss minimization (relative)	objectives: cost minimization	objectives: cutting cost minimization	objectives: others	status of information: deterministic	status of information: stochastic	variability: fixed data	variability: variable data	solution approaches – pattern-orientated: single-patterned	solution approaches – pattern-orientated: multi-patterned
Haessler (1980a)	●				●	●					●		●			●					●		●			●
Heicken/König (1980)			●		●	●					●		●		●		●		●		●			●		●
Fang/Lamendola (1982)	●				●		●				●			●		●			●		●		●			●
Karmarkar/Karp (1982)	●				●		●				●		●					●			●		●			●
Chvatal (1983)	●				●	●					●		●		●	●					●		●			●
Johnston (1986)			●		●	●					●		●			●					●			●		●
Olorunniwo (1986)	●				●	●					●		●		●						●		●			●
Diegel (1988a)		●			●	●					●		●		●						●		●			●
Haessler (1988b)-1		●			●	●				●	●			●				●	●		●			●		●
Haessler (1988b)-2		●			●	●				●		●		●				●	●		●			●		●

Haessler (1988b)-3		●			●	●				●	●		●					●			●		●			●
Haessler (1988b)-4		●			●	●					●		●					●	●		●			●		●
Haessler (1988b)-5		●			●	●					●		●					●	●		●			●		●
Stadtler (1988)	●				●	●					●		●					●			●		●			●
Wäscher (1989a)-1	●				●	●					●		●		●						●		●			●
Wäscher (1989a)-2	●				●	●					●		●		●						●		●			●
Wäscher (1989a)-3	●				●	●					●		●		●				●	● 2)	●		●			●
Stadtler (1990)			●		●		●	●			●		●		●						●			●		●

2) trim loss removal cost minimization
left over cost minimization
overrun cost minimization

Table 7.4b: CS1-problems with discrete quantity measurement and a homogeneous assortment of large objects (Part 2)

7.1.2.2. Discrete CS1-Type with a Heterogeneous Assortment of Large Objects

(1) Common characteristics

The common characteristics found in sources of CS1-types with discrete quantity measurement and a heterogeneous assortment of large objects are signed in Table 7.5. In all cases the heterogeneous assortment of the objects is fixed. Therefore, this property is marked as type-defining.

Common characteristics found aside from those type-defining ones are the following:

- fixed figures of items,
- discrete quantity measurement of small items,
- limited number of items per figure,
- no restrictions concerning the pattern sequence,
- pattern-orientated methods.

(2) Comparative contrast of literary sources

Those contributions treating CS1-types with discrete quantity measurement and a heterogeneous measurement of large objects are presented in the following tables:

- Table 7.6a: Without pattern restrictions, Part 1
- Table 7.6b: Without pattern restrictions, Part 2
- Table 7.6c: Only pattern restrictions, Part 1
- Table 7.6d: Only pattern restrictions, Part 2

The following ascertainments were made from the investigation:

- Two-thirds of the studies are either application-orientated or case studies, the latter of which represents 63%.

- The available number of large objects appears in all possible formations. Mostly (86%), there is an unlimited number per figure; a mixed number per figure only appears in 8% of the studies.

- In 94% of the contributions there are no restrictions in object sequence. In the two studies in which there is a certain order, one is partial (Eng/Daellenbach 1985) and one is complete (Fisk/Hung 1979).

- The sequence of small items is predominantly optional (86%). An example of a partial order appears in Fisk/Hung (1979); a complete order appears in Eilon (1960), Tokuyama/Ueno (1985), Sumichrast (1986) and Sweeney/Haessler (1990).

- Distancing restrictions and frequency limitations play a role in only 19% of the studies in which a distance to the edge and/or restrictions to the number or to the combination of items are to be considered. Separation restrictions are not valid in one-dimensional problems.

- Predominantly single-stage processes are employed in the problems (83%).

- Restrictions in the number of patterns only appear in 8% of the contributions.

- There is often (64%) only one objective in the process, namely input minimization, relative or absolute trim loss minimization, material or space cost minimization or value maximization. In many cases change-over (cutting) and/or inventory costs are of importance, and in one case transport costs. One case transforms various objectives into an overall "utility" function, maximizing profit contribution (Atkins et al. 1984).

- A majority of the problems have a deterministic status of information.

- Fixed data are given in two-thirds of the sources.

- 83% of the solution procedures employ pattern-orientated methods.

attributes			properties			
dimensionality			1-DIMENSIONAL	2-dimensional	3-dimensional	
type of assignment			Type II	TYPE III	TYPE IV	
large objects	type of quantity measurement		discrete	continuous		
	figure	form	—	—		
	assortment	homogeneous	fixed	variable		
		heterogeneous	FIXED	variable		
	availability	number/figure	1	limited	unlimited	mixed
		sequence	no order	partial order	complete order	
small items	type of quantity measurement		discrete	continuous		
	figure	form	—	—		
		orientation	fixed	optional	90° turn (parallel)	90° turn (optional)
	assortment	homogeneous	fixed	variable		
		heterogeneous	fixed	variable	few objects per figure	MANY OBJECTS PER FIGURE
	availability	number/figure	limited	unlimited		
		sequence	no order	partial order	complete order	
		dates	identical time references	different time references		
pattern restrictions	geometric characteristics	distancing restrictions	none	between items	to the edges	
		frequency limitations	none	limited combinations of figures	limited number of figures	limited number of items
		separation restrictions	—	—	—	—
	operational characteristics	number of stages	single-stage processes	multi-stage processes		
		pattern sequence	no restrictions	with restrictions		
		pattern frequency	no restrictions	limited number of different patterns	minimum number of identical patterns	
	dynamics of allocation		dynamic	static: off-line	static: on-line	
objectives			input minimization	trim loss minimization	value maximization	others
status of information			deterministic	stochastic		
variability of data			fixed data	variable data		
solution methods			object-orientated methods	single-patterned methods	multi-patterned methods	

Table 7.5: Common characteristics of the CS1-type with discrete quantity measurement and a heterogeneous assortment of large objects

properties	type of sources			large objects					small items				objectives								status of information		variability		solution approaches	
				availability					availability																pattern-orientated	
				number/figure			sequence		sequence		dates															
sources / author(s) and year of publication	theoretical/methodic	application-orientated	case study	limited	unlimited	mixed	no order	order	no order	order	identical time reference	different time reference	input minimization	trim loss minimization (absolute)	trim loss minimization (relative)	cost minimization	cutting cost minimization	value maximization	inventory cost minimization	others	deterministic	stochastic	fixed data	variable data	single-patterned	multi-patterned
Atkins et al. (1984)			●		●		●		●		●									● 1)	●			●		●
Eng/Daellenbach (1985)		●		●				●	●		●							●			●		●			●
Tokuyama/Ueno (1985)			●		●		●			●	●				●							●		●		●
Wäscher et al. (1985)			●	●			●		●			●		●			●		●		●			●	●	
Roodman (1986)	●			●			●		●		●			●							●		●			●
Seth et al. (1986)			●	●			●		●		●				●						●		●			●
Sumichrast (1986)			●		●		●			●	●					●	●				●			●		●
Wäscher/Müller (1986)			●			●	●		●		●				●		●		●		●			●		●
Farley (1988c)	●					●	●		●		●					●			●		●			●		●
Ferreira et al. (1990)		●			●		●		●		●				●		●				●			●	●	
Sweeney/Haessler (1990)		●		●			●			●	●							●			●		●			●

1) maximization of profit contribution

Table 7.6a: CS1-problems with discrete quantity measurement and a heterogeneous assortment of large objects (without pattern restrictions, Part 1)

properties / sources: author(s) and year of publication	type of sources: theoretical/methodic	type of sources: application-orientated	type of sources: case study	large objects: availability: number/figure: limited	large objects: availability: number/figure: unlimited	large objects: availability: number/figure: mixed	large objects: availability: sequence: no order	large objects: availability: sequence: order	small items: availability: sequence: no order	small items: availability: sequence: order	small items: availability: dates: identical time reference	small items: availability: dates: different time reference	objectives: input minimization	objectives: trim loss minimization (absolute)	objectives: trim loss minimization (relative)	objectives: cost minimization	objectives: cutting cost minimization	objectives: value maximization	objectives: inventory cost minimization	objectives: others	status of information: deterministic	status of information: stochastic	variability: fixed data	variability: variable data	solution approaches: pattern-orientated: single-patterned	solution approaches: pattern-orientated: multi-patterned
Paull (1956)			●		●		●		●		●			●							●		●			●
Eisemann (1957)		●			●		●		●		●					●					●		●			●
Eilon (1960)-2			●	●			●			●	●			●								●	●			●
Gilmore/Gomory (1961)	●				●		●		●		●					●					●		●			●
Gilmore/Gomory (1963)	●				●		●		●		●					●					●		●			●
Pierce (1964)		●			●		●		●		●					●					●		●			●
Bernhard (1967)	●			●			●		●		●					●			●		●		●			●
Beged-Dov (1970)			●		●		●		●		●			●						● 1)	●		●			●
Lasdon (1972)-1	●				●		●		●		●		●								●		●			●
Lasdon (1972)-2	●				●		●		●		●			●							●			●		●
Johnston/Bourke (1973)		●			●		●		●		●					●					●			●		●
Haessler (1975)	●				●		●		●		●			●			●				●			●		●
Coffield/Crisp (1976)		●		●			●		●			●		●			●		●		●		●			●

Rao (1976)	●				●		●		●		●					●					●		●			●
Kallio (1977)			●		●		●		●		●			●							●		●		●	●
Litton (1977)		●			●		●		●			●			●								●		●	
Stainton (1977)			●			●	●		●		●			●							●		●			●
Hung/Brown (1978)	●			●			●		●		●		●								●		●		●	
Fisk/Hung (1979)	●				●			●		●	●					●					●		●		●	
Gaudioso (1979)	●				●		●		●		●		●								●		●			●
Haessler/Vonderembse (1979)			●	●			●		●		●			●		●	●				●			●		●
Vonderembse (1979)			●	●			●		●		●			●		●	●				●			●		●
Sculli (1981)			●		●		●		●		●			●								●	●		●	
Tokuyama/Ueno (1981)			●	●			●		●		●				●						●		●			●

1) minimization of transportation costs

Table 7.6b: CS1-problems with discrete quantity measurement and a heterogeneous assortment of large objects (without pattern restrictions, Part 2)

properties / sources	pattern restrictions																		
	geometry									organization						dynamics of allocation			
	distance		frequency			separation				number of stages		sequence		frequency of patterns		static		dynamic	
							orthogonal												
								guillotine patterns											
author(s) and year of publication	between items	to the edges	limited combinations of figures	limited number of figures	limited number of items	non-orthogonal	nested patterns	limited stages	unlimited stages	single-stage	multi-stage	unrestricted	restricted	limited number of different pattern types	minimum number of identical patterns	on-line	off-line	with realization	without realization
Paull (1956)										●		●					●		
Eisemann (1957)										●		●					●		
Eilon (1960)-2										●		●					●		
Gilmore/Gomory (1961)										●		●					●		
Gilmore/Gomory (1963)										●		●					●		
Pierce (1964)										●		●					●		
Bernhard (1967)										●		●					●		
Beged-Dov (1970)										●		●					●		
Lasdon (1972)-1										●		●					●		

Lasdon (1972)-2										●		●					●		
Johnston/Bourke (1973)					●					●		●					●		
Haessler (1975)										●		●					●		
Coffield/Crisp (1976)										●		●					●		
Rao (1976)											●	●					●		
Kallio (1977)										●		●					●		
Litton (1977)										●		●							●
Stainton (1977)										●		●					●		
Hung/Brown (1978)										●		●					●		
Fisk/Hung (1979)										●		●					●		
Gaudioso (1979)										●		●					●		
Haessler/Vonderembse (1979)											●		●						
Vonderembse (1979)											●		●						

Table 7.6c: CS1-problems with discrete quantity measurement and a heterogeneous assortment of large objects (only pattern restrictions, Part 1)

properties	pattern restrictions																		
	geometry									organization						dynamics of allocation			
	distance		frequency			separation				number of stages		sequence		frequency of patterns		static		dynamic	
							orthogonal												
								guillotine patterns											
sources / author(s) and year of publication	between items	to the edges	limited combinations of figures	limited number of figures	limited number of items	non-orthogonal	nested patterns	limited stages	unlimited stages	single-stage	multi-stage	unrestricted	restricted	limited number of different pattern type	minimum number of identical patterns	on-line	off-line	with reallocation	without reallocation
Sculli (1981)										●		●					●		
Tokuyama/Ueno (1981)										●		●					●		
Atkins et al. (1984)											●	●					●		
Eng/Daellenbach (1985)										●		●					●		
Tokuyama/Ueno (1985)			●							●		●				●			
Wäscher et al. (1985)		●			●						●		●				●		

Roodman (1986)										●		●					●		
Seth et al. (1986)										●		●					●		
Sumichrast (1986)										●		●					●		
Wäscher/Müller (1986)		●								●		●					●		
Farley (1988c)										●		●					●		
Ferreira et al. (1990)		●									●	●					●		
Sweeney/Haessler (1990)		●	●							●		●					●		

Table 7.6d: CS1-problems with discrete quantity measurement and a heterogeneous assortment of large objects (only pattern restrictions, Part 2)

7.2. Two-dimensional Cutting Stock Types (CS2)

7.2.1 CS2-Type with Non-rectangular Small Items

(1) Common characteristics

Table 7.7 identifies the common characteristics of the CS2-type with non-rectangular items. Aside from the type-defining characteristics are the following similarities:

- discrete quantity measurement of objects and items,
- rectangular large objects,
- fixed assortments of objects and items,
- no given order of large objects,
- limited number of small items per figure,
- small items of identical time reference,
- non-orthogonal patterns,
- no limitations in regard to pattern frequency,
- (static) off-line problems,
- pattern-orientated methods.

(2) Comparative contrast of literary sources

The sources for the CS2-type with non-rectangular items are described in Table 7.8a (without pattern restrictions) and in Table 7.8b (only pattern restrictions). For reasons of editorial simplification the common properties 'non-orthogonal' and 'off-line' are identified once again in Table 7.8b.

Dominating the matrices are the following characteristics:

- application-orientated studies or case studies (85%),
- a heterogeneous assortment of large objects (65%),
- no given order of objects (95%),
- unlimited number of objects (70%),
- no given order of small items (75%),
- single-stage processes (95%),
- optional pattern sequence (95%),
- single-goal problems (65%),
- deterministic status of information (90%),
- fixed data (90%).

attributes			properties			
dimensionality			1-dimensional	2-DIMENSIONAL	3-dimensional	
type of assignment			Type II	TYPE III	TYPE IV	
large objects	type of quantity measurement		discrete	continuous		
	figure	form	rectangular	non-rectangular		
	assortment	homogeneous	fixed	variable		
		heterogeneous	fixed	variable		
	availability	number/figure	1	limited	unlimited	mixed
		sequence	no order	partial order	complete order	
small items	type of quantity measurement		discrete	continuous		
	figure	form	rectangular	NON-RECTANGULAR		
		orientation	fixed	optional	90° turn (parallel)	90° turn (optional)
	assortment	homogeneous	fixed	variable		
		heterogeneous	fixed	variable	few objects per figure	MANY OBJECTS PER FIGURE
	availability	number/figure	limited	unlimited		
		sequence	no order	partial order	complete order	
		dates	identical time references	different time references		
pattern restrictions	geometric characteristics	distancing restrictions	none	between items	to the edges	
		frequency limitations	none	limited combinations of figures	limited number of figures	limited number of items
		separation restrictions	non-orthogonal patterns	nested patterns	guillotine patterns (limited stages)	guillotine patterns (unlimited stages)
	operational characteristics	number of stages	single-stage processes	multi-stage processes		
		pattern sequence	no restrictions	with restrictions		
		pattern frequency	no restrictions	limited number of different patterns	minimum number of identical patterns	
	dynamics of allocation		dynamic	static: off-line	static: on-line	
objectives			input minimization	trim loss minimization	value maximization	others
status of information			deterministic	stochastic		
variability of data			fixed data	variable data		
solution methods			object-orientated methods	single-patterned methods	multi-patterned methods	

Table 7.7: Common characteristics of the CS2-type with non-rectangular small items

properties / sources: author(s) and year of publication	type of sources: theoretical/methodic	type of sources: application-orientated	type of sources: case study	large objects: assortment: heterogeneous	large objects: assortment: homogeneous	large objects: availability number/figure: limited	large objects: availability number/figure: unlimited	large objects: availability number/figure: mixed	small items: shape orientation: fixed	small items: shape orientation: any	small items: shape orientation: 90 degree-turns (parallel)	small items: availability sequence: no order	small items: availability sequence: partial order	small items: availability sequence: complete order	objectives: input minimization	objectives: trim loss minimization (absolute)	objectives: trim loss minimization (relative)	objectives: cost minimization	objectives: others	status of information: deterministic	status of information: stochastic	variability: fixed data	variability: variable data	solution approaches (pattern-orientated): single-patterned	solution approaches (pattern-orientated): multi-patterned
Schachtel (1958)		●			●	●						●					●	●	● 1)	●		●		●	
Lampl/Stahl (1965)			●		●	●				●		●				●				●		●			●
Adamowicz/Albano (1972)-2		●			●	●				●		●					●			●		●		●	
Steuckart (1974)-1		●		●			●		●				●				●		● 2)	●		●			
Steuckart (1974)-2		●		●			●			●			●				●		● 2)	●		●			
Steuckart (1974)-3		●		●			●				●		●				●		● 2)	●		●			
Albano (1977)		●			●	●				●		●				●				●		●		●	
Eversheim/Herngesberg (1977)		●		●			●		●			●				●				●		●			●
Albano/Sapuppo (1980)	●				●		●			●		●				●				●		●		●	
Ellinger et al. (1980)			●	●			●		●			●						●		●		●			●
Troßmann (1983)		●			●		●		●			●						●		●			●		
Israni/Sanders (1984)		●		●			●					●			●					●		●		●	

Roberts (1984)-1		●		●			●		●			●				●			●[3]	●		●			●
Roberts (1984)-2		●		●			●				●	●				●			●[3]	●		●			●
Meier/Naß (1986)			●	●			●							●			●			●		●			
Stommel/Buschhoff (1986)			●	●				●				●					●			●		●			
Dagli/Tatoglu (1987)	●			●		●				●		●			●					●		●			●
Farley (1988b)		●		●			●		●				●						●[4]		●		●		●
Qu/Sanders (1989)	●			●			●					●					●			●		●		●	●
Farley (1990a)		●			●		●					●				●					●	●			●

1) working cost minimization
tool cost minimization
2) cutting cost minimization,
working time minimization,
working cost minimization
3) minimization of the number of cuts
4) net return maximization

Table 7.8a: CS2-problems with non-rectangular small items (without pattern restrictions)

properties / sources	pattern restrictions: geometry									pattern restrictions: organization						dynamics of allocation			
	distance		frequency			separation				number of stages		sequence		frequency of patterns		static		dynamic	
							orthogonal												
								guillotine patterns											
author(s) and year of publication	between items	to the edges	limited combinations of figures	limited number of figures	limited number of items	non-orthogonal	nested patterns	limited stages	unlimited stages	single-stage	multi-stage	unrestricted	restricted	different pattern types limited number of	minimum number of identical patterns	on-line	off-line	with reallocation	without reallocation
Schachtel (1958)	●	●				●				●		●					●		
Lampl/Stahl (1965)						●				●		●					●		
Adamowicz/Albano (1972)-2						●				●		●					●		
Steuckart (1974)-1	●	●				●				●		●					●		
Steuckart (1974)-2	●	●				●				●		●					●		
Steuckart (1974)-3	●	●				●				●		●					●		
Albano (1977)	●					●				●		●					●		
Eversheim/Hemgesberg (1977)	●	●				●				●		●					●		
Albano/Sapuppo (1980)						●				●		●					●		

Ellinger et al. (1980)						●				●		●					●		
Troßmann (1983)						●											●		
Israni/Sanders (1984)						●				●		●					●		
Roberts (1984)-1						●				●		●					●		
Roberts (1984)-2						●				●		●					●		
Meier/Naß (1986)			●			●					●		●				●		
Stommel/Buschhoff (1986)						●				●		●					●		
Dagli/Tatoglu (1987)	●					●				●		●					●		
Farley (1988b)						●				●		●					●		
Qu/Sanders (1989)						●				●		●					●		
Farley (1990a)						●				●		●					●		

Table 7.8b: CS2-problems with non-rectangular small items (only pattern restrictions)

7.2.2 CS2-Types with Rectangular Small Items

7.2.2.1 Rectangular CS2-Types with Only One Large Object per Figure

In this type C&P problems are unique in that a given number of small rectangles with certain lengths and widths must be assigned to a large object of given width in such a way as to minimize waste of "length". For the purpose of this investigation the situation is idealized so that, rather than taking an object of infinite length, infinite ones of this width and all possible lengths are presented, from which the shortest serves for the assignment.

(1) Common characteristics

Table 7.9 displays these common characteristics of the CS2-type with rectangular small items and one large object per figure:

- discrete quantity measurement of objects and items,
- rectangular large objects,
- heterogeneous, fixed assortment of objects,
- no given order of large objects,
- fixed assortment of small items,
- limited number of small items per figure,
- small items of identical time reference,
- no distancing restrictions or frequency limitations,
- single-stage processes,
- no restrictions concerning pattern sequence or frequency,
- (static) off-line problems,
- deterministic status of information
- single-patterned methods.

attributes			properties			
dimensionality			1-dimensional	2-DIMENSIONAL	3-dimensional	
type of assignment			Type II	TYPE III	TYPE IV	
large objects	type of quantity measurement		discrete	continuous		
	figure	form	rectangular	non-rectangular		
	assortment	homogeneous	fixed	variable		
		heterogeneous	fixed	variable		
	availability	number/figure	1	limited	unlimited	mixed
		sequence	no order	partial order	complete order	
small items	type of quantity measurement		discrete	continuous		
	figure	form	RECTANGULAR	non-rectangular		
		orientation	fixed	optional	90° turn (parallel)	90° turn (optional)
	assortment	homogeneous	fixed	variable		
		heterogeneous	fixed	variable	few objects per figure	MANY OBJECTS PER FIGURE
	availability	number/figure	limited	unlimited		
		sequence	no order	partial order	complete order	
		dates	identical time references	different time references		
pattern restrictions	geometric characteristics	distancing restrictions	none	between items	to the edges	
		frequency limitations	none	limited combinations of figures	limited number of figures	limited number of items
		separation restrictions	non-orthogonal patterns	nested patterns	guillotine patterns (limited stages)	guillotine patterns (unlimited stages)
	operational characteristics	number of stages	single-stage processes	multi-stage processes		
		pattern sequence	no restrictions	with restrictions		
		pattern frequency	no restrictions	limited number of different patterns	minimum number of identical patterns	
	dynamics of allocation		dynamic	static: off-line	static: on-line	
objectives			input minimization	trim loss minimization	value maximization	others
status of information			deterministic	stochastic		
variability of data			fixed data	variable data		
solution methods			object-orientated methods	single-patterned methods	multi-patterned methods	

Table 7.9: Common characteristics of the CS2-type with rectangular small items and only one large object per figure

(2) Comparative contrast of literary sources

As represented in Table 7.10, the following observations can be made about the CS2-type mentioned above:

- There are predominantly theoretical/methodical studies (86%).
- There is largely a given assortment of large objects (86%).
- Small objects are mostly not given a certain order.
- Orthogonal patterns dominate.
- Mostly there is only one objective, which is to minimize input or absolute trim loss (71%).
- There are normally fixed data (71%).

It is significant that most of the deviations from these standard properties are found in the only application-orientated study (Moll 1985). Other studies of particular interest are those by De Cani (1978) and Rinnoy Kan et al. (1978), which deal with non-orthogonal patterns (for rectangular small objects).

properties	type of sources		small items							pattern restrictions				objectives				variability	
			shape of figures			assortment		availability		geometry									
			orientation			heterogeneous		sequence		separation									
											orthogonal								
												guillotine patterns							
sources / author(s) and year of publication	theoretical/methodic	application-orientated	fixed	any	90 degree-turns (parallel)	fixed	variable	no order	partial order	non-orthogonal	nested patterns	limited stages	unlimited stages	input minimization	trim loss minimization (absolute)	cutting cost minimization	others	fixed data	variable data
Filmer (1970)	●					●			●				●		●	●			●
De Cani (1978)-1	●					●		●		●				●				●	
De Cani (1978)-2	●				●	●		●			●			●				●	
Biro/Boros (1984)	●		●			●		●			●			●				●	
Moll (1985)		●					●	●					●		●		● 1)		●
Chauny et al. (1987)	●				●	●		●			●			●				●	
Rinnooy Kan et al. (1987)	●			●		●		●		●					●			●	

1) trim loss goals, capacity goals, goals of satisfying demand, timing goals, clipping goals, alternative-production goals

Table 7.10: CS2-problems with rectangular small items and only one object per figure

7.2.2.2 Rectangular CS2-Types with Guillotine Patterns

(1) Common characteristics

The common characteristics of the CS2-type with rectangular small items and guillotine patterns are identified in Table 7.11. Aside from the type-defining characteristics are the following common traits:

- discrete quantity measurement of objects and items,
- fixed assortments of objects and items,
- no given order of large objects,
- limited number of small items per figure,
- small objects with identical time reference,
- (static) off-line problems,
- deterministic status of information,
- pattern-orientated methods.

attributes			properties			
dimensionality			1-dimensional	2-DIMENSIONAL	3-dimensional	
type of assignment			Type II	TYPE III	TYPE IV	
large objects	type of quantity measurement		discrete	continuous		
	figure	form	rectangular	non-rectangular		
	assortment	homogeneous	fixed	variable		
		heterogeneous	fixed	variable		
	availability	number/figure	1	limited	unlimited	mixed
		sequence	no order	partial order	complete order	
small items	type of quantity measurement		discrete	continuous		
	figure	form	RECTANGULAR	non-rectangular		
		orientation	fixed	optional	90° turn (parallel)	90° turn (optional)
	assortment	homogeneous	fixed	variable		
		heterogeneous	fixed	variable	few objects per figure	MANY OBJECTS PER FIGURE
	availability	number/figure	limited	unlimited		
		sequence	no order	partial order	complete order	
		dates	identical time references	different time references		
pattern restrictions	geometric characteristics	distancing restrictions	none	between items	to the edges	
		frequency limitations	none	limited combinations of figures	limited number of figures	limited number of items
		separation restrictions	non-orthogonal patterns	nested patterns	GUILLOTINE PATTERNS (limited stages)	GUILLOTINE PATTERNS (unlimited stages)
	operational characteristics	number of stages	single-stage processes	multi-stage processes		
		pattern sequence	no restrictions	with restrictions		
		pattern frequency	no restrictions	limited number of different patterns	minimum number of identical patterns	
	dynamics of allocation		dynamic	static: off-line	static: on-line	
objectives			input minimization	trim loss minimization	value maximization	others
status of information			deterministic	stochastic		
variability of data			fixed data	variable data		
solution methods			object-orientated methods	single-patterned methods	multi-patterned methods	

Table 7.11: Common characteristics of the CS2-type with rectangular small items and guillotine patterns

(2) Comparative contrast of literary sources

The following tables compare sources of the CS2-type with rectangular small items and guillotine patterns:

- Table 7.12a: Without pattern restrictions (Part 1)
- Table 7.12b: Without pattern restrictions (Part 2)
- Table 7.12c: Only pattern restrictions (Part 1)
- Table 7.12d: Only pattern restrictions (Part 2)

After an overall examination of these tables, the following conclusions arise:

- The sources are divided rather equally between theoretical/methodical studies and application-orientated or case studies.

- Large objects have rectangular forms in 97% of the cases.

- The assortment of large objects is generally heterogeneous (86%).

- Two-thirds of the contributions (64%) treated problems with heterogeneous assortments of the large objects.

- Concerning the available number of large objects, all three possible properties are represented, although the most common (72%) is an unlimited number per figure.

- In a third of the studies (33%) there are restrictions in the sequence of small objects.

- In the majority of the cases (78%) there is only one objective where the goal is to minimize input, trim loss (absolute or relative) or cost of material or space. Other objectives do appear, however, particularly in the studies after 1983 (Table 7.12a).

- There are fixed data in 83% of the studies.

- Pattern-orientated methods were employed for 75% of the contributions.

Almost all of the studies typified as the above mentioned CS2-type differ in reference to at least one of the decision-relevant properties. This leads to the conclusion that most sources concerning this type each represents a number of actual C&P problems and can thus be considered as problem type themselve.

properties	type of sources			large objects							small items				objectives							variability		solution approaches	
				form		assortment		availability			shape		availability											pattern-orientated	
								number/figure			orientation		sequence												
sources / author(s) and year of publication	theoretical/methodic	application-orientated	case study	rectangular	non-rectangular	heterogeneous	homogeneous	limited	unlimited	mixed	fixed	90 degree-turns (parallel)	no order	partial order	input minimization	trim loss minimization (absolute)	trim loss minimization (relative)	cost minimization	cutting cost minimization	value maximization	others	fixed data	variable data	single-patterned	multi-patterned
Fabian (1984)	●			●			●		●				●			●						●			●
Farley/Richardson (1984)		●		●			●		●				●					●	●			●			●
Harrison (1984)		●		●		●			●				●			●					● 1)	●			●
Beasley (1985c)	●			●		●		●			●		●					●		●		●		●	
Schneider (1987)	●			●			●		●		●		●			●						●		●	●
Seth (1987)			●		●	●			●			●	●					●				●			●
Madsen (1988)			●	●		●			●			●		●		●					● 2)	●			●
Albat/Wirsam (1989)		●		●		●							●					●	●				●		
Vasko et al. (1989)		●		●		●		●				●		●							● 3)	●			●
Farley (1990b)-1	●			●		●		●			●		●				●					●			●
Farley (1990b)-2	●			●		●		●				●	●				●					●			●

1) volume throughput minimization, pattern type minimization, order spread minimization
2) order spread minimization
3) net return maximization

Table 7.12a: CS2-problems with rectangular small items and guillotine patterns (without pattern restrictions, Part 1)

sources: author(s) and year of publication	type of sources: theoretical/methodic	type of sources: application-orientated	type of sources: case study	large objects: form: rectangular	large objects: form: non-rectangular	large objects: assortment: heterogeneous	large objects: assortment: homogeneous	large objects: availability: number/figure: limited	large objects: availability: number/figure: unlimited	large objects: availability: number/figure: mixed
Pegels (1967b)		●		●		●			●	
Hahn (1968)	●			●		●			●	
Wihl (1969)-1			●	●		●			●	
Wihl (1969)-2			●	●		●			●	
Heim (1971)		●		●		●		●		
Adamowicz/Albano (1972)		●		●			●	●		
Gerhardt (1973)	●			●		●		●		
Gribov (1973)	●			●			●		●	
Dyson/Gregory (1974)			●	●		●			●	
Page (1975)	●			●		●		●		
Adamowicz/Albano (1976a)	●			●		●			●	
Skalbeck/Schultz (1976)	●			●		●		●		

sources: author(s) and year of publication	small items: shapes: orientation: fixed	small items: shapes: orientation: 90 degree-turns (parallel)	small items: availability: sequence: no order	small items: availability: sequence: partial order	objectives: input minimization	objectives: trim loss minimization (absolute)	objectives: trim loss minimization (relative)	objectives: cost minimization	objectives: cutting cost minimization	objectives: value maximization	objectives: others
Pegels (1967b)				●			●		●		
Hahn (1968)		●	●				●				
Wihl (1969)-1	●		●				●				
Wihl (1969)-2		●	●					●			
Heim (1971)			●			●					
Adamowicz/Albano (1972)			●				●				
Gerhardt (1973)			●			●					
Gribov (1973)		●	●				●				
Dyson/Gregory (1974)				●		●					
Page (1975)	●		●					●			
Adamowicz/Albano (1976a)		●	●		●						
Skalbeck/Schultz (1976)		●	●				●				

sources: author(s) and year of publication	variability: fixed data	variability: variable data	solution approaches: pattern-orientated: single-patterned	solution approaches: pattern-orientated: multi-patterned
Pegels (1967b)		●		●
Hahn (1968)	●		●	
Wihl (1969)-1		●		●
Wihl (1969)-2		●		●
Heim (1971)	●		●	
Adamowicz/Albano (1972)	●		●	
Gerhardt (1973)	●		●	
Gribov (1973)	●			●
Dyson/Gregory (1974)	●			●
Page (1975)	●		●	
Adamowicz/Albano (1976a)	●		●	
Skalbeck/Schultz (1976)	●			●

Hiroxman (1977)		●		●		●			●			●		●		●						●			●
Meier (1978)			●	●		●				●	●			●			●						●	●	●
Madsen (1979)			●	●			●		●			●	●			●						●			●
Albano/Orsini (1980b)-1	●			●		●			●		●			●		●						●		●	
Albano/Orsini (1980b)-2	●			●		●			●			●		●		●						●		●	
Rijckaert (1980)	●			●			●		●		●		●					●				●			●
Ellinger et al. (1981)			●	●			●		●			●	●			●						●			●
Mangalousis et al. (1981)		●		●		●			●				●			●						●			●
Rijckaert (1981)	●			●			●		●			●	●					●				●			●
Laubenstein et al. (1982)			●	●		●			●				●		●	●							●		
Smithin/Harrison (1982)		●		●		●			●				●			●								●	
Farley (1983a)		●		●			●		●				●		●	●						●			●
Farley (1983b)	●			●			●		●				●		●	●						●			●
Harrison (1983)		●		●			●	●									●	●	●						●
Wang (1983)	●			●			●		●		●		●			●						●			●
Duta/Fabian (1984)	●			●		●				●			●			●						●			●

Table 7.12b: CS2-problems with rectangular small items and guillotine patterns (without pattern restrictions, Part 2)

sources / properties: author(s) and year of publication	pattern restrictions – geometry – distance: between items	pattern restrictions – geometry – distance: to the edges	pattern restrictions – geometry – frequency: limited combinations of figures	pattern restrictions – geometry – frequency: limited number of figures	pattern restrictions – geometry – frequency: limited number of items	pattern restrictions – geometry – separation: non-orthogonal	pattern restrictions – geometry – separation – orthogonal: nested patterns	pattern restrictions – geometry – separation – orthogonal – guillotine patterns: limited stages	pattern restrictions – geometry – separation – orthogonal – guillotine patterns: unlimited stages	pattern restrictions – organization – number of stages: single-stage	pattern restrictions – organization – number of stages: multi-stage	pattern restrictions – organization – sequence: unrestricted	pattern restrictions – organization – sequence: restricted	pattern restrictions – organization – frequency of patterns: limited number of different patterns	pattern restrictions – organization – frequency of patterns: minimum number of identical patterns	dynamics of allocation – static: on-line	dynamics of allocation – static: off-line	dynamics of allocation – dynamic: with reallocation	dynamics of allocation – dynamic: without reallocation
Pegels (1967b)											●	●					●		
Hahn (1968)			●							●		●					●		
Wihl (1969)-1										●		●					●		
Wihl (1969)-2										●		●					●		
Heim (1971)										●		●					●		
Adamowicz/Albano (1972)									●	●		●					●		
Gerhardt (1973)										●		●					●		
Gribov (1973)									●	●		●					●		
Dyson/Gregory (1974)			●		●					●			●				●		

Page (1975)									●		●					●		
Adamowicz/Albano (1976a)			●					●	●		●					●		
Skalbeck/Schultz (1976)									●		●					●		
Hinxman (1977)		●			●				●		●					●		
Meier (1978)			●						●			●				●		
Madsen (1979)									●		●					●		
Albano/Orsini (1980b)-1								●	●		●					●		
Albano/Orsini (1980b)-2								●	●		●					●		
Rijckaert (1980)								●	●		●					●		
Ellinger et al. (1981)								●		●	●					●		
Mangalousis et al. (1981)										●	●					●		
Rijckaert (1981)								●	●		●					●		

Table 7.12c: CS2-problems with rectangular small items and guillotine patterns (pattern restrictions, Part 1)

properties	pattern restrictions																
	geometry									organization						dynamics of allocation			
	distance		frequency			separation				number of stages		sequence		frequency of patterns		static		dynamic	
							orthogonal												
								guillotine pattern											
sources / author(s) and year of publication	between items	to the edges	limited combinations of figures	limited number of figures	limited number of items	non-orthogonal	nested patterns	limited stages	unlimited stages	single-stage	multi-stage	unrestricted	restricted	limited number of different patterns	minimum number of identical patterns	on-line	off-line	with reallocation	without reallocation
Laubenstein et al. (1982)					●						●		●				●		
Smithin/Harrison (1982)									●	●		●					●		
Farley (1983a)										●		●					●		
Farley (1983b)										●		●					●		
Harrison (1983)										●		●					●		
Wang (1983)									●	●		●					●		
Dutta/Fabian (1984)									●	●		●					●		
Fabian (1984)										●		●					●		
Farley/Richardson (1984)										●			●				●		

Harrison (1984)										●		●		●			●		
Beasley (1985c)									●	●		●					●		
Schneider (1987)									●	●		●					●		
Seth (1987)									●	●		●					●		
Madsen (1988)										●			●				●		
Albat/Wirsam (1989)										●		●					●		
Vasko et al. (1989)									●	●		●					●		
Farley (1990b)-1										●		●					●		
Farley (1990b)-2										●		●					●		

Table 7.12d: CS2-problems with rectangular small items and guillotine patterns (only pattern restrictions, Part 2)

7.2.2.3 Rectangular CS2-Type with Nested Patterns

(1) Common characteristics

Table 7.13 displays the CS2-type with rectangular small items and nested patterns. The following similarities appear apart from the type-defining characteristics:

- discrete quantity measurement of objects and items,
- rectangular large objects,
- fixed assortments of objects and items,
- no given order of large objects,
- a permitted 90° rotation of small items,
- limited number of items per figure,
- small items of identical time reference,
- no distancing restrictions or frequency limitations,
- no restrictions in pattern sequence,
- (static) off-line problems,
- deterministic status of information,
- fixed data,
- pattern-orientated methods.

(2) Comparative contrast of literary sources

There are only four contributions for the CS2-type with rectangular small items and nested patterns as shown in Table 7.14. These problems differ at most in their objectives.

attributes			properties			
dimensionality			1-dimensional	2-DIMENSIONAL	3-dimensional	
type of assignment			Type II	TYPE III	TYPE IV	
large objects	type of quantity measurement		discrete	continuous		
	figure	form	rectangular	non-rectangular		
	assortment	homogeneous	fixed	variable		
		heterogeneous	fixed	variable		
	availability	number/figure	1	limited	unlimited	mixed
		sequence	no order	partial order	complete order	
small items	type of quantity measurement		discrete	continuous		
	figure	form	RECTANGULAR	non-rectangular		
		orientation	fixed	optional	90° turn (parallel)	90° turn (optional)
	assortment	homogeneous	fixed	variable		
		heterogeneous	fixed	variable	few objects per figure	MANY OBJECTS PER FIGURE
	availability	number/figure	limited	unlimited		
		sequence	no order	partial order	complete order	
		dates	identical time references	different time references		
pattern restrictions	geometric characteristics	distancing restrictions	none	between items	to the edges	
		frequency limitations	none	limited combinations of figures	limited number of figures	limited number of items
		separation restrictions	non-orthogonal patterns	NESTED PATTERNS	guillotine patterns (limited stages)	guillotine patterns (unlimited stages)
	operational characteristics	number of stages	single-stage processes	multi-stage processes		
		pattern sequence	no restrictions	with restrictions		
		pattern frequency	no restrictions	limited number of different patterns	minimum number of identical patterns	
	dynamics of allocation		dynamic	static: off-line	static: on-line	
objectives			input minimization	trim loss minimization	value maximization	others
status of information			deterministic	stochastic		
variability of data			fixed data	variable data		
solution methods			object-orientated methods	single-patterned methods	multi-patterned methods	

Table 7.13: Common characteristics of the CS2-type with rectangular small items and nested patterns

properties	type of sources		large objects				small items		pattern restrictions			objectives					
			assortment		availability		availability		organization								
					number/figure		sequence		number of stages		frequency of patterns						
sources / author(s) and year of publication	theoretical/methodic	case study	heterogeneous	homogeneous	limited	unlimited	no order	partial order	single-stage	multi-stage	limited number of different patterns	trim loss minimization (absolute)	trim loss minimization (relative)	cost minimization	cutting cost minimization	inventory cost minimization	productivity maximization
Tokuyama/Ueno (1981)-2		●	●		●		●			●			●				
Bleibohm/Kuhse (1982)		●	●			●	●		●			●		●	●	●	
Fabian/Duta (1982)	●		●			●	●		●		●	●					●
Toczylowski (1986)	●			●		●		●	●					●			

Table 7.14: CS2-problems with rectangular small items and nested patterns

7.3 Three-dimensional Cutting Stock Type (CS3)

(1) Common characteristics

The common characteristics found in sources of the CS3-type are presented in Table 7.15. Apart from the type-defining characteristics are the following similarities:

- discrete quantity measurement of objects and items,
- rectangular objects and items,
- no given order of large objects,
- heterogeneous, fixed assortments of objects and items,
- limited number of small items per figure,
- small items with identical time reference,
- single-stage processes,
- no restrictions for pattern sequence or frequency,
- (static) off-line problems,
- deterministic status of information,
- fixed data,
- pattern-orientated methods.

(2) Comparative contrast of literary sources

Only five contributions handle CS3-types. Their comparative analysis is displayed in Table 7.16.

attributes			properties			
dimensionality			1-dimensional	2-dimensional	3-DIMENSIONAL	
type of assignment			Type II	TYPE III	TYPE IV	
large objects	type of quantity measurement		discrete	continuous		
	figure	form	rectangular	non-rectangular		
	assortment	homogeneous	fixed	variable		
		heterogeneous	fixed	variable		
	availability	number/figure	1	limited	unlimited	mixed
		sequence	no order	partial order	complete order	
small items	type of quantity measurement		discrete	continuous		
	figure	form	rectangular	non-rectangular		
		orientation	fixed	optional	90° turn (parallel)	90° turn (optional)
	assortment	homogeneous	fixed	variable		
		heterogeneous	fixed	variable	few objects per figure	MANY OBJECTS PER FIGURE
	availability	number/figure	limited	unlimited		
		sequence	no order	partial order	complete order	
		dates	identical time references	different time references		
pattern restrictions	geometric characteristics	distancing restrictions	none	between items	to the edges	
		frequency limitations	none	limited combinations of figures	limited number of figures	limited number of items
		separation restrictions	non-orthogonal patterns	nested patterns	guillotine patterns (limited stages)	guillotine patterns (unlimited stages)
	operational characteristics	number of stages	single-stage processes	multi-stage processes		
		pattern sequence	no restrictions	with restrictions		
		pattern frequency	no restrictions	limited number of different patterns	minimum number of identical patterns	
	dynamics of allocation		dynamic	static: off-line	static: on-line	
objectives			input minimization	trim loss minimization	value maximization	others
status of information			deterministic	stochastic		
variability of data			fixed data	variable data		
solution methods			object-orientated methods	single-patterned methods	multi-patterned methods	

Table 7.15: Common characteristics of the CS3-type

properties / sources: author(s) and year of publication	type of sources: theoretical/methodic	type of sources: application-orientated	type of sources: case study	large objects: availability: number/figure: limited	large objects: availability: number/figure: unlimited	small items: shape: orientation: fixed	small items: shape: orientation: 90 degree-turns (parallel)	small items: shape: orientation: 90 degree-turns (any)	small items: availability: sequence: no order	small items: availability: sequence: partial order	pattern restrictions: geometry: distance to the edges	pattern restrictions: fre-quency: limited combinations of figures	pattern restrictions: separation: orthogonal: nested patterns	pattern restrictions: separation: orthogonal: guillotine patterns: limited stages	pattern restrictions: separation: orthogonal: guillotine patterns: unlimited stages	objectives: input minimization	objectives: trim loss minimization (absolute)	objectives: cutting cost minimization	objectives: others	solution approaches: pattern-orientated: single-patterned	solution approaches: pattern-orientated: multi-patterned
Kortanek/Sodaro (1966)	●			●					●								●	●	● 1)		●
Merle/Grütz (1978)			●		●			●		●		●	●						● 2)		●
Schneider (1979)	●			●		●			●			●		●			●				●
Liu/Chen (1981)		●		●					●				●			●				●	
Schneider (1988)			●	●		●			●		●			●			●			●	●

1) working cost minimization
2) delivery cost minimization

Table 7.16: CS3-problems

7.4 Actual Cutting Stock Problems

The properties of reality-based attributes of sources for CS-types are presented in the following tables:

(1) **CS1-type with continuous quantity measurement of large objects (Table 7.17):**

- The application-orientated sources or case studies of this type exclusively deal with cutting problems belonging to the paper (58%), iron and steel (27%), textile (11%) and plastic industries (9%).

- 39% of the contributions isolate the cutting problem. In the other studies there are at least one or two connections to other planning areas, mostly with scheduling and/or inventory planning.

- Software for practical application exists in 35% of the analysed problems.

(2) **CS1-type with discrete quantity measurement and a homogeneous assortment of large objects (Table 7.18):**

- The contributions for this type also exclusively deal with cutting problems. The branch orientation is similar to the type with continuous quantity measurement.

- With respect to the planning context the study by Heicken/König (1980) emphasizes many interdependencies to other areas of production planning. In other sources there are relatively few considered connections to additional areas of planning.

- From what could be determined from the sources, software for practical application exists for about half of the given problems.

(3) **CS1-type with discrete quantity measurement and a heterogeneous assortment of large objects (Table 7.19):**

- The contributions which apply to this type deal with cutting problems, 39% belong to the iron and steel industries and 26% to the paper industry. The remaining cutting problems are in the areas of carpet sales (1), timber industry (2), forge (1), weaving mill (1) or of no specific area (1).

- Two-thirds (63%) of the sources deal with cutting problems in connection with other areas of planning, in which sequencing is almost always employed.

- Software solutions provide practical application for a third of the problems described.

(4) **CS2-type with non-rectangular small items (Table 7.20):**

- Sources of this type deal with cutting problems, primarily within the textile industry (35%) or the iron and steel industries (24%). The remaining problems are equally represented either in the ship building, furniture or canvas industries, or in no specific branch of industry.

- A significant 71% of the contributions combine cutting problems with other areas of planning. Particularly frequent are order management capacity planning, scheduling, choice of manufacturing method and inventory planning.

- 30% of the contributions provide practical software solutions.

(5) **CS2-type with rectangular small items (Table 7.21):**

The matrix representing reality-based characteristics the CS2-type with rectangular small items is divided into three parts according to the developed hierarchy of types (one large object per figure, guillotine patterns, nested patterns). From an overview of the matrix these observations can be made:

- The contributions for this type exclusively deal with cutting problems namely in the areas of glass (7), iron and steel (5), timber (5), furniture (3), paper (2), ship building (1) or they are not specified to any area (1).

- One half of the sources approach the cutting problem in an isolated fashion and the other half mostly in reference to sequencing.

- Software for practical application appears in a third of the problems.

(6) **CS3-type (Table 7.22):**

- Of the contributions for the CS3-type, two problems arise from mail order selling or container loading and one is a cutting problem from the rubber processing industry.

- The two packing problems are treated in connection with packaging planning or order processing, whereas the cutting problem is isolated.

- Applicable software is available for the packing problem in mail order selling.

properties / sources: author(s) and year of publication	branch of industry	kind of objects and items: large objects	kind of objects and items: small items	planning situation – cutting problems: isolated cutting planning	with order mangement	with timing	with capacity planning	with sequencing	with choice of manufacturing methods	with numerical control	with inventory planning	with material requirement planning	with procurement planning	others	planning situation – packing problems: isolated packing planning	with packaging planning	others	software: practical application	simulation
Vajda (1958)	paper industry	rolls of paper	rolls of paper	●															
Förstner (1959)-1	paper industry	rolls of paper	rolls of paper	●														●	
Förstner (1959)-2	paper industry	rolls of paper	rolls of paper								●							●	
Förstner (1963)	paper industry	rolls of paper	rolls of paper				●				●								
Meerendonk (1963)	paper industry	rolls of paper	rolls of paper	●															
Poirier (1967)	paper industry	rolls of paper	rolls of paper			●		●	●									●	
Caruso/Kokat (1973)	iron and steel industry	rolled metal	rolled metal		●			●						with transport planning				●	
Goswami (1973)	iron and steel industry	rolled metal	transformer sheet								●							●	
Beged-Dov (1974)	paper industry	rolls of paper	rolls of paper	●															●
Hardley (1976)	iron and steel industry	steel coils	steel coils					●										●	
Haessler (1978)-1	iron and steel industry	steel coils	steel coils	●															●
Haessler (1978)-2	iron and steel industry	steel coils	steel coils	●															●

Schepens (1978)	paper industry	corrugated cardboard	corrugated cardboard				●		●									●	
Haessler (1979)	plastics industry	plastic film	plastic film					●											●
Dyckhoff (1981)	textile industry	rolls of material	length of material	●															●
Dyckhoff/Gehring (1982)	textile industry	rolls of synthetic cloth	rolls of synthetic cloth								●								●
Johnston (1982)	paper industry	rolls of paper	rolls of paper					●											●
Vonderembse/Haessler (1982)	iron and steel industry	steel slabs	steel slabs					●										●	
Haessler/Talbot (1983)	paper industry	rolls of paper	rolls of paper						●										●
Haessler (1983)	paper industry	rolls of paper	rolls of paper	●															●
Bartmann (1986)	paper industry	rolls of paper	rolls of paper		●	●		●		●	●							●	
Tabucanon/Lertcharoen-sombat (1986)	paper industry	rolls of paper	rolls of paper	●															●
Dyckhoff/Gehring (1988)	textile industry	rolls of synthetic cloth	rolls of synthetic cloth								●								●
Nickels (1988)	paper industry	rolls of paper	rolls (pieces) of paper			●	●	●	●		●								●
Goulimis (1990)	paper industry	rolls of paper	rolls of paper	●															●
Wäscher (1990)	iron and steel industry	steel coils	steel coils								●								●

Table 7.17: Actual CS1-problems with continuous quantity measurement of large objects

properties / sources: author(s) and year of publication	branch of industry	kind of objects and items: large objects	kind of objects and items: small items	planning situation – cutting planning: isolated cutting planning	with order management	with timing	with capacity planning	with sequencing	with choice of manufacturing methods	with numerical control	with inventory planning	with material requirement planning	with procurement planning	others	planning situation – packing problems: isolated packing planning	with packaging planning	others	software: practical application	simulation
Metzger (1958)	iron and steel industry	steel coils	steel coils	●															●
Eilon (1960)-1	iron and steel industry	steel bars	steel bars						●										●
Kreko (1965)	paper industry	rolls of paper	rolls of paper	●															
Johns (1967)	paper industry	rolls of paper	rolls of paper	●															●
Haessler (1971)	paper industry	rolls of paper	rolls of paper						●										●
Coverdale/Wharton (1976)	paper industry	rolls of paper	rolls of paper					●	●										●
Ahluwalia/Saxena (1978)	iron and steel industry	steel slabs	steel slabs								●	●	●						●
Heicken/König (1980)	iron and steel industry	tubes	tubes		●	●	●	●	●	●	●	●	●					●	
Johnston (1986)	paper industry	rolls of paper	rolls of paper	●														●	

Diegel (1988e)	paper industry	rolls of paper	rolls of paper	●														●	
Haessler (1988b)-1	paper industry	rolls of paper	rolls of paper					●											
Haessler (1988b)-2	paper industry	rolls of paper	rolls of paper					●											
Haessler (1988b)-3	paper industry	corrugated cardboard	corrugated cardboard						●									●	
Haessler (1988b)-4	plastics industry	plastic film	plastic film					●										●	
Haessler (1988b)-5	paper industry	rolls of paper	rolls of paper					●										●	
Stadtler (1990)	iron and steel industry	aluminium profiles	aluminium profiles	●														●	

Table 7.18: Actual CS1-problems with discrete quantity measurement and a homogeneous assortment of large objects

properties / sources: author(s) and year of publication	branch of industry	kind of objects and items: large objects	kind of objects and items: small items	planning situation: cutting problems: isolated cutting planning	with order management	with timing	with capacity planning	with sequencing	with choice of manufacturing methods	with numerical control	with inventory planning	with material requirement planning	with procurement planning	others	planning situation: packing problems: isolated packing planning	with packaging planning	others	software: practical application	simulation
Paull (1956)	paper industry	rolls of paper	rolls of paper	●															
Eisemann (1957)		rolls of material	rolls of material	●															●
Eilon (1960)-2	iron and steel industry	steel bars	steel bars	●															●
Pierce (1964)	paper industry	rolls of paper	rolls of paper	●															●
Beged-Dov (1970)	paper industry	rolls of paper	rolls of paper				●							with transport planning					
Johnston/Bourke (1973)	paper industry	rolls of paper	rolls of paper						●									●	
Coffield/Crisp (1976)		rolls of material	rolls of material			●	●	●	●	●	●	●	●					●	
Kallio (1977)	paper industry	rolls of paper	rolls of paper		●			●										●	
Litton (1977)	carpet industry	carpet rolls	pieces of carpet								●								●
Stainton (1977)	iron and steel industry	steel reinforcement bars	steel reinforcement bars					●			●								●
Haessler/Vonderembse (1979)	iron and steel industry	steel slabs	steel slabs			●		●											●
Vonderembse (1979)	iron and steel industry	steel slabs	steel slabs			●		●											●

Sculli (1981)		rolls of isolated tape	rolls of isolated tape						●										
Tokuyama/Ueno (1981)	iron and steel industry	steel plates	steel plates					●	●		●							●	
Atkins et al. (1984)	timber industry	blocks of wood	pieces of chipboard	●										with assortment selection problem				●	
Eng/Daellenbach (1985)	timber industry	stems	short logs	●															●
Tokuyama/Ueno (1985)	iron and steel industry	steel blooms	steel blooms				●	●		●	●								●
Wäscher et al. (1985)	iron and steel industry	steel coils	steel coils		●	●		●	●	●	●							●	
Seth et al. (1986)	forge shop	steel bars	steel bars	●														●	
Sumnichrast (1986)	weaving mill	rolls of woven goods	rolls of woven goods	●														●	
Wäscher/Müller (1986)	iron and steel industry	steel plates	steel plates					●			●								●
Ferreira et al. (1990)	iron and steel industry	rolls of material	rolls of material				●												●
Sweeney/Haessler (1990)	paper industry	rolls of paper	rolls of paper					●											●

Table 7.19: Actual CS1-problems with discrete quantity measurement and a heterogeneous assortment of large objects

properties: sources / author(s) and year of publication	branch of industry	kind of objects and items: large objects	kind of objects and items: small items	planning situation – cutting problems: isolated cutting planning	with order management	with timing	with capacity planning	with sequencing	with choice of manufacturing systems	with numerical control	with inventory planning	with material requirement planning	with procurement planning	others	planning situation – packing problems: isolated packing problem	with packaging planning	others	software: practical application	simulation
Schachtel (1958)	iron and steel industry	metal sheets	pieces of sheet metal	●															●
Lampl/Stahl (1965)		rectangular plates	circles	●															●
Adamowicz/Albano (1972)-2	shipbuilding	steel slabs	steel slabs	●															●
Steuckart (1974)-1	textile industry	length of material	parts of trousers				●		●										
Steuckart (1974)-2	textile industry	length of material	parts of trousers				●		●										
Steuckart (1974)-3	textile industry	length of material	parts of trousers				●		●										
Albano (1977)	shipbuilding	metals of steel	metals of steel	●															●
Eversheim/Hemgesberg (1977)	iron and steel industry	metal sheets	pieces of sheet metal							●									●
Ellinger et al. (1980)	textile industry	length of material	parts of trousers					●										●	
Troßmann (1983)	textile industry	length of material	pieces of clothes					●											
Israni/Sanders (1984)	iron and steel industry	steel slabs	pieces of steel							●	●								

Roberts (1984)-1	furniture industry	rectangular boards	worktops		●		●				●							●	
Roberts (1984)-2	furniture industry	rectangular boards	worktops		●		●				●							●	
Meier/Naß (1986)	plastics industry	plastic sheets	pieces of plastic		●			●		●								●	
Stommel/Buschhoff (1986)	iron and steel industry	metal sheets	pieces of metal sheet		●	●		●		●								●	
Farley (1988b)	textile industry	length of material	pieces of material				●												●
Farley (1990a)	canvas industry	length of canvas	tents etc.	●															●

Table 7.20: Actual CS2-problems with non-rectangular small items

	properties		kind of objects and items		planning situation														software	
					cutting problems											packing problems				
sources	author(s) and year of publication	branch of industry	large objects	small items	isolated cutting planning	with order management	with timing	with capacity planning	with sequencing	with choice of manufacturing methods	with numerical control	with inventory planning	with material requirement planning	with procurement planning	others	isolated packing planning	with packaging planning	others	practical application	simulation
1 large object/figure	Moll (1985)	paper industry	rolls of corrugated paper	folding boxes			●	●	●											●
guillotine patterns	Pegels (1967b)	paper industry	corrugated paper	corrugated paper					●	●										●
	Wihl (1969)-1	timber industry	pieces of chipboard	pieces of chipboard	●															●
	Wihl (1969)-2	timber industry	pieces of chipboard	pieces of chipboard	●															●
	Heim (1971)		plates	plates								●	●	●						●
	Adamowicz/Albano (1972)-1	shipbuilding	steel slabs	steel slabs	●															●
	Dyson/Gregory (1974)	glass industry	glass plates	glass plates					●				●							●
	Hinxman (1977)	glass industry	glass plates	glass plates	●															●
	Meier (1978)	iron and steel industry	rolled metals	rolled metals		●		●	●										●	
	Madson (1979a)	glass industry	glass plates	isolated glass plates					●											●
	Ellinger et al. (1981)	glass industry	glass plates	isolated glass plates	●														●	
	Mangalousis et al. (1981)	timber industry	boards	panels	●															●
	Laubenstein et al. (1982)	iron and steel industry	metal sheets	metal sheets					●										●	

guillotine patterns	Smithin (1982)	furniture industry	boards	panels	●														●	
	Farley (1983a)	glass industry	glass plates in stock	glass plates as ordered	●														●	
	Harrison (1983)	furniture industry	boards	panels	●															
	Farley/Richardson (1984)	glass industry	glass plates	glass plates					●											●
	Harrison (1984)	furniture industry	pieces of chipboard	panels				●	●											
	Seth (1987)	timber industry	rectangular parallelopiped	planks	●															●
	Madsen (1988)	glass industry	glass plates	glass plates		●			●											●
	Albat/Wirsam (1989)	timber industry	pieces of chipboard	pieces of chipboard	●														●	
	Vasko et al. (1989)	iron and steel industry	metal sheets	pieces of metal sheet	●															●
nested patterns	Tokuyama/Ueno (1981)-2	iron and steel industry	steel plates	steel plates								●							●	
	Bleibohm/Kuhse (1982)	iron and steel industry	metal sheets	metal sheets			●		●		●	●	●	●					●	

Table 7.21: Actual CS2-problems with rectangular small items

properties / sources		kind of objects and items		planning situation														software	
				cutting problems											packing problems				
author(s) and year of publication	branch of industry	large objects	small items	isolated cutting planning	with order management	with timing	with capacity planing	with sequencing	with choice of manufacturing methods	with numerical control	with inventory planning	with material requirement planning	with procurement planning	others	isolated packing planning	with packaging planning	others	practical application	simulation
Merle/Grütz (1978)	mail order business	packets	articles for dispatch													●		●	
Liu/Chen (1981)		container	boxes														with order management		●
Schneider (1988)	rubber industry	crepe rubber	crepe rubber	●															●

Table 7.22: Actual CS3-problems

8. Knapsack Types (KS)

8.1 One-dimensional Knapsack Type (KS1)

(1) Common characteristics

The common characteristics of the KS1-type are presented in Table 8.1. Aside from the type-defining characteristics are the following similarities:

- discrete quantity measurement of objects and items,
- limited numbers of objects and items,
- fixed assortments of objects and items,
- small items with identical time reference,
- no restrictions concerning pattern sequence or frequency,
- object-orientated or single patterned methods.

(2) Comparative contrast of literary sources

The following properties appear most frequently in the matrix (one should note that there are only nine problems of this type):

- theoretical/methodical studies (78%),
- identical measurements: a homogeneous assortment of large objects (67%),
- several large objects (67%),
- no given order of large objects (67%),
- no geometric characteristics (89%),
- single-stage processes (89%),
- off-line problems (78%),
- value maximization (89%),
- deterministic status of information (89%),
- fixed data (86%).

There are studies with exactly these "standard properties" (e.g. Bruno/Downey 1985), as well as some in which deviations of one or more properties (e.g. Lee 1979) exist.

attributes			properties			
dimensionality			1-DIMENSIONAL	2-dimensional	3-dimensional	
type of assignment			TYPE II	Type III	Type IV	
large objects	type of quantity measurement		discrete	continuous		
	figure	form	—	—		
	assortment	homogeneous	fixed	variable		
		heterogeneous	fixed	variable		
	availability	number/figure	1	limited	unlimited	mixed
		sequence	no order	partial order	complete order	
small items	type of quantity measurement		discrete	continuous		
	figure	form	—	—		
		orientation	—	—	—	—
	assortment	homogeneous	fixed	variable		
		heterogeneous	fixed	variable	FEW OBJECTS PER FIGURE	MANY OBJECTS PER FIGURE
	availability	number/figure	limited	unlimited		
		sequence	no order	partial order	complete order	
		dates	identical time references	different time references		
pattern restrictions	geometric characteristics	distancing restrictions	none	between items	to the edges	
		frequency limitations	none	limited combinations of figures	limited number of figures	limited number of items
		separation restrictions	—	—	—	—
	operational characteristics	number of stages	single-stage processes	multi-stage processes		
		pattern sequence	no restrictions	with restrictions		
		pattern frequency	no restrictions	limited number of different patterns	minimum number of identical patterns	
	dynamics of allocation		dynamic	static: off-line	static: on-line	
objectives			input minimization	trim loss minimization	value maximization	others
status of information			deterministic	stochastic		
variability of data			fixed data	variable data		
solution methods			object-orientated methods	single-patterned methods	multi-patterned methods	

Table 8.1: Common characteristics of the KS1-type

sources: author(s) and year of publication / properties	type of sources: theoretical/methodic	type of sources: case study	large objects – assortment: heterogeneous	large objects – assortment: homogeneous	large objects – availability – number/figure: one large object	large objects – availability – number/figure: several large objects	large objects – availability – sequence: no order	large objects – availability – sequence: complete order	small items – availability – sequence: no order	small items – availability – sequence: partial order	small items – availability – sequence: complete order	pattern restrictions – geometry – distance: between items	pattern restrictions – geometry – frequency: limited combinations of figures	pattern restrictions – geometry – frequency: limited number of objects	pattern restrictions – organization – number of stages: single-stage	pattern restrictions – organization – number of stages: multi-stage	pattern restrictions – dynamics of allocation – static: on-line	pattern restrictions – dynamics of allocation – static: off-line	objectives: trim loss minimization (relatively)	objectives: value maximization	status of information: deterministic	status of information: stochastic	solution approaches: item-orientated	solution approaches – pattern-orientated: single-patterned	solution approaches – pattern-orientated: multi-patterned
Tilanus/Gerhardt (1976)		●	●			●		●			●	●				●	●			●		●		●	
Cheng/Pila (1978)	●		●			●	●		●						●			●	●		●			●	
Coffman et al. (1978b)	●			●		●	●		●						●			●		●	●		●		
Fisk/Hung (1979)-2	●		●			●		●			●				●			●		●	●			●	
Lee (1979)		●		●	●		●			●			●	●	●			●		●	●				
Bruno/Downey (1985)	●			●		●	●		●						●			●		●	●		●		
Sarker (1988)	●			●	●		●		●						●			●		●	●			●	
Foster/Vohra (1989)-1	●			●		●	●		●						●			●		●	●		●		
Foster/Vohra (1989)-2	●			●	●		●		●						●		●			●		●	●		

Table 8.2: KS1-problems

8.2 Two-dimensional Knapsack Type (KS2)

(1) Common characteristics

Table 8.3 presents the common characteristics existing in sources for the KS2-type. Aside from the type-defining characteristics are the following similarities:

- discrete quantity measurement of objects and items,
- one large object per figure,
- fixed assortments of objects and items,
- no given order of objects or items,
- rectangular items,
- small items of identical time reference,
- no distancing restrictions or frequency limitations,
- single-stage processes,
- no restrictions concerning pattern sequence or frequency,
- deterministic status of information,
- fixed data,
- object- or pattern-orientated methods.

(2) Comparative contrast of literary sources

Table 8.4 presents a comparative contrast of literary sources for KS2-type of which these are the most common characteristics:

- theoretical/methodical studies (95%),
- rectangular large objects (95%),
- nested patterns (57%),
- value maximization (91%),
- single-patterned solution methods (95%).

attributes			properties			
dimensionality			1-dimensional	2-DIMENSIONAL	3-dimensional	
type of assignment			TYPE II	Type III	Type IV	
large objects	type of quantity measurement		discrete	continuous		
	figure	form	rectangular	non-rectangular		
	assortment	homogeneous	fixed	variable		
		heterogeneous	fixed	variable		
	availability	number/figure	1	limited	unlimited	mixed
		sequence	no order	partial order	complete order	
small items	type of quantity measurement		discrete	continuous		
	figure	form	rectangular	non-rectangular		
		orientation	fixed	optional	90° turn (parallel)	90° turn (optional)
	assortment	homogeneous	fixed	variable		
		heterogeneous	fixed	variable	FEW OBJECTS PER FIGURE	MANY OBJECTS PER FIGURE
	availability	number/figure	limited	unlimited		
		sequence	no order	partial order	complete order	
		dates	identical time references	different time references		
pattern restrictions	geometric characteristics	distancing restrictions	none	between items	to the edges	
		frequency limitations	none	limited combinations of figures	limited number of figures	limited number of items
		separation restrictions	non-orthogonal patterns	nested patterns	guillotine patterns (limited stages)	guillotine patterns (unlimited stages)
	operational characteristics	number of stages	single-stage processes	multi-stage processes		
		pattern sequence	no restrictions	with restrictions		
		pattern frequency	no restrictions	limited number of different patterns	minimum number of identical patterns	
	dynamics of allocation		dynamic	static: off-line	static: on-line	
objectives			input minimization	trim loss minimization	value maximization	others
status of information			deterministic	stochastic		
variability of data			fixed data	variable data		
solution methods			object-orientated methods	single-patterned methods	multi-patterned methods	

Table 8.3 Common characteristics of the KS2-type

properties / sources	type of sources		small items						pattern restrictions						objectives			solution approaches	
			form		orientation		availability number/figure		separation				number of stages						
										orthogonal									
											guillotine patterns								
author(s) and year of publication	theoretical/methodic	application-oriented	rectangular	non-rectangular	fixed	90 degree-turns (parallel)	unlimited	limited	non-orthogonal	nested patterns	limited stages	unlimited stages	single-stage	multi-stage	trim loss minimization (absolute)	value maximization	trim loss minimization (relative)	pattern-oriented (one pattern)	item-oriented
Brooks et al. (1940)	●		●		●			●		●			●			●		●	
Barnett/Kynch (1967)	●		●					●		●			●			●		●	
Romanovskii (1969)	●		●		●			●				●	●			●		●	
Haims/Freeman (1970)	●			●	●			●	●				●			●		●	
Christofides/Whitlock (1977)	●		●		●			●				●	●			●		●	
Barnes (1979)	●		●			●	●			●			●			●		●	
Hodgson (1982)-1	●		●			●		●		●			●			●		●	
Laurent/Iyengar (1982)	●		●			●		●		●			●			●		●	
Baker/Calderbank (1983)	●		●			●		●		●			●			●			●

Hodgson et al. (1983)	●		●			●	●			●			●			●		●	
Lam (1983)	●		●					●		●			●				●	●	
Beasley (1985d)	●		●		●		●				●		●			●		●	
Beasley (1985a)	●		●		●			●		●			●			●		●	
Beasley (1985b)	●		●		●			●		●			●			●		●	
Zissimopoulos (1985)	●		●		●		●					●	●			●		●	
Penington/Tanchoco (1988)		●	●			●	●					●	●			●		●	
Scheithauer/Terno (1988c)	●		●			●		●				●	●			●		●	
Tsai et al. (1988)	●		●			●		●		●			●			●		●	
Vasko (1989)	●		●			●		●				●	●			●		●	
Dowsland (1990)	●		●			●		●		●			●			●		●	
Oliveira/Ferreira (1990)	●		●		●			●				●	●		●			●	

Table 8.4 KS2-problems

8.3 Three-dimensional Knapsack Type (KS3)

(1) Common characteristics

Common characteristics arising from the sources on the KS3-type are identified in Table 8.5. Aside from those type-defining characteristics are the following commonalities:

- discrete quantity measurement of objects and items,
- limited number of large objects,
- fixed assortments of objects and items,
- rectangular small items,
- single-stage processes,
- no restrictions concerning pattern sequence or frequency,
- fixed data,
- object- or pattern-orientated methods.

attributes			properties			
dimensionality			1-dimensional	2-dimensional	3-DIMENSIONAL	
type of assignment			TYPE II	Type III	Type IV	
large objects	type of quantity measurement		discrete	continuous		
	figure	form	rectangular	non-rectangular		
	assortment	homogeneous	fixed	variable		
		heterogeneous	fixed	variable		
	availability	number/figure	1	limited	unlimited	mixed
		sequence	no order	partial order	complete order	
small items	type of quantity measurement		discrete	continuous		
	figure	form	rectangular	non-rectangular		
		orientation	fixed	optional	90° turn (parallel)	90° turn (optional)
	assortment	homogeneous	fixed	variable		
		heterogeneous	fixed	variable	FEW OBJECTS PER FIGURE	MANY OBJECTS PER FIGURE
	availability	number/figure	limited	unlimited		
		sequence	no order	partial order	complete order	
		dates	identical time references	different time references		
pattern restrictions	geometric characteristics	distancing restrictions	none	between items	to the edges	
		frequency limitations	none	limited combinations of figures	limited number of figures	limited number of items
		separation restrictions	non-orthogonal patterns	nested patterns	guillotine patterns (limited stages)	guillotine patterns (unlimited stages)
	operational characteristics	number of stages	single-stage processes	multi-stage processes		
		pattern sequence	no restrictions	with restrictions		
		pattern frequency	no restrictions	limited number of different patterns	minimum number of identical patterns	
	dynamics of allocation		dynamic	static: off-line	static: on-line	
objectives			input minimization	trim loss minimization	value maximization	others
status of information			deterministic	stochastic		
variability of data			fixed data	variable data		
solution methods			object-orientated methods	single-patterned methods	multi-patterned methods	

Table 8.5 Common characteristics of the KS3-type

(2) Comparative contrast of literary sources

Contributions for the KS3-type are arranged and compared in two tables, those with no pattern restrictions (Table 8.6a) and those only for the pattern restrictions (Table 8.6b). Although the examined studies are limited to only eight, the following points should still be considered:

- 86% of the studies are application-orientated or case studies.
- The assortments of large objects are divided fairly equally between heterogeneous and homogeneous groups.
- It was not possible to determine the orientation of small items in all cases. In those studies which identify the orientation, it is either fixed or a permitted 90-degree turns (parallel or any).
- Mostly, there is a limitation in number of small items per figure (86%).
- No limitations are stated concerning the sequence of small items in 57% of the studies.
- Distancing restrictions or frequency limitations appear in only two cases.
- The patterns are generally orthogonal.
- The majority (86%) of the problems are off-line.
- Problems with small items of identical time reference dominate (86%).
- With the exception of one case, there is only one objective, whether it be (absolute) trim loss minimization or value maximization. In the multi-criteria solution by Sculli/Hui (1988) various objectives are integrated in one utility function.
- The status of information is normally deterministic (85,7%).
- There are dominantly single-patterned procedures.

properties	type of sources			large objects						small items										objectives			status of information		solution approaches	
				form		assortment		availability		shape			availability													pattern-oriented
								number/figure		orientation			number/figure		sequence			dates								
								limited																		
sources / author(s) and year of publication	theoretical/methodic	application-oriented	case study	rectangular	non-rectangular	heterogeneous	homogeneous	one large object	limited number	fixed	90 degree-turns (parallel)	90 degree-turns (any)	unlimited	limited	no order	partial order	complete order	identical time references	different time references	trim loss minimization (absolutely)	value maximization	utility maximization	deterministic	stochastic	item-oriented	single-patterned
Hodgson (1982)-2	●			●			●	●				●		●	●			●			●		●			●
Faaland/Briggs (1984)		●			●	●		●					●		●			●			●		●			●
Geerts (1984)		●			●	●			●	●				●	●			●			●		●			●
Sculli/Hui (1988)		●		●			●	●		●				●			●		●			●		●	●	
Reinders/Hendriks (1989)			●		●	●			●					●	●			●			●		●			●
Gehring et al. (1990)		●		●			●	●				●		●			●	●		●			●			●
Haessler/Talbot (1990)			●	●		●		●			●			●		●		●		●			●			●

Table 8.6a: KS3-problems (without pattern restrictions)

properties	pattern restrictions																		
	geometry									organization						dynamics of allocation			
	distance		frequency			separation				number of stages		sequence		frequency of patterns		static		dynamic	
							orthogonal												
								guillotine patterns											
sources / author(s) and year of publication	between items	to the edges	limited combinations of figures	limited number of figures	limited number of items	non-orthogonal	nested patterns	limited stages	unlimited stages	single-stage	multi-stage	unrestricted	restricted	limited number of different pattern types	minimum number of identical patterns	on-line	off-line	with reallocation	without reallocation
Hodgson (1982)-2							●			●		●					●		
Faaland/Briggs (1984)								●		●		●					●		
Geerts (1984)							●			●		●					●		
Sculli/Hui (1988)			●					●		●		●						●	
Reinders/Hendriks (1989)									●	●		●					●		
Gehring et al. (1990)							●			●		●					●		
Haessler/Talbot (1990)	●						●			●		●					●		

Table 8.6b: KS3-problems (only pattern restrictions)

8.4 Actual Knapsack Problems

Reality-based characteristics found in application-orientated studies or case studies concerning the knapsack type are presented in Table 8.7. The sources are classified according to dimensionality, refering to the hierarchy of types.

- Of the two KS1-type problems, one is a cutting problem in the steel industry and the other a packing problem in loading tankers. The former is handled in coordination with time-scheduling, the latter is isolated. Applicable software is available for the packing problem.

- In the single KS2-type application-orientated study, there is a problem of piling freight containers on to pallets. This problem is handled in an isolated fashion and includes applicable software.

- The contributions of KS3-types are divided rather equally between cutting and packing problems. All of the cutting problems involve the isolated planning of timbering. In the packing problems, loading ship containers and vehicles or optimally distributing storage space are considered either in connection with stability considerations and scheduling or in an isolated fashion. The only study employing applicable software is in loading vehicles by Haessler/Talbot (1990).

sources	properties: author(s) and year of publication	branch of industry	kind of objects and items: large objects	kind of objects and items: small items	planning situation – cutting problems: isolated cutting planning	with order management	with timing	with capacity planning	with sequencing	with choice of manufacturing methods	with numerical control	with inventory planning	with material requirement planning	with procurement planning	others	planning situation – packing problems: isolated packing planning	with packaging planning	others	software: practical application	software: simulation
1-dimensional	Tilanus/Gerhardt (1976)	iron and steel industry	steel slabs	steel plates			●													●
1-dimensional	Lee (1979)		tank lorries	tanks												●			●	
2-dimensional	Penington/Tanchoco (1988)		pallets	cargo boxes												●			●	
3-dimensional	Faaland/Briggs (1984)	timber industry	tree-trunks	lumber	●															●
3-dimensional	Geerts (1984)		logs	short logs	●															●
3-dimensional	Sculli/Hui (1988)		storage space	ship containers												●				●
3-dimensional	Reinders/Hendriks (1989)	timber industry	logs	lumber	●															●
3-dimensional	Gehring et al. (1990)		ship containers	cargo boxes														with balancing		●
3-dimensional	Haessler/Talbot (1990)		vehicles	cargo boxes														with sequencing	●	

Table 8.7: Actual knapsack problems

9. Pallet Loading Types (PL)

9.1 Two-dimensional Pallet Loading Type (PL2)

(1) Common characteristics

Table 9.1 provides the list of the common characteristics of the PL2-type. Aside from the type-defining properties are the following similarities:

- discrete quantity measurement of objects and items,
- rectangular large objects,
- homogeneous assortment of large objects,
- one large object per figure,
- no order of items and objects,
- small items with identical time reference,
- no frequency restrictions,
- single-stage processes,
- no restrictions concerning pattern sequencing or frequency,
- (static) off-line problems,
- deterministic status of information,
- fixed data,
- single-patterned methods.

attributes			properties			
dimensionality			1-dimensional	2-DIMENSIONAL	3-dimensional	
type of assignment			TYPE II	Type III	Type IV	
large objects	type of quantity measurement		discrete	continuous		
	figure	form	rectangular	non-rectangular		
	assortment	homogeneous	fixed	variable		
		heterogeneous	fixed	variable		
	availability	number/figure	1	limited	unlimited	mixed
		sequence	no order	partial order	complete order	
small items	type of quantity measurement		discrete	continuous		
	figure	form	rectangular	non-rectangular		
		orientation	fixed	optional	90° turn (parallel)	90° turn (optional)
	assortment	homogeneous	FIXED	variable		
		heterogeneous	fixed	variable	few objects per figure	many objects per figure
	availability	number/figure	limited	unlimited		
		sequence	no order	partial order	complete order	
		dates	identical time references	different time references		
pattern restrictions	geometric characteristics	distancing restrictions	none	between items	to the edges	
		frequency limitations	none	limited combinations of figures	limited number of figures	limited number of items
		separation restrictions	non-orthogonal patterns	nested patterns	guillotine patterns (limited stages)	guillotine patterns (unlimited stages)
	operational characteristics	number of stages	single-stage processes	multi-stage processes		
		pattern sequence	no restrictions	with restrictions		
		pattern frequency	no restrictions	limited number of different patterns	minimum number of identical patterns	
	dynamics of allocation		dynamic	static: off-line	static: on-line	
objectives			input minimization	trim loss minimization	value maximization	others
status of information			deterministic	stochastic		
variability of data			fixed data	variable data		
solution methods			object-orientated methods	single-patterned methods	multi-patterned methods	

Table 9.1: Common characteristics of the PL2-type

(2) Comparative contrast of literary sources

The contributions of the PL2-type are contrasted in Table 9.2, with these properties dominating:

- fixed measurements of large objects (80%),
- rectangular small items (95%),
- a permitted 90° rotation of small items (95%),
- fixed measurements of small items (85%),
- limited number of small items per figure (85%),
- no distancing restrictions (95%),
- nested patterns (90%),
- value maximization (85%), often with stability optimization (41%).

The most deviations of these standard properties are in the study by Dorri/Ben-Basset (1984) in which the small items exist in irregularly-shaped forms. It should be further noted that the sources of this type generally handle application-orientated studies which discuss pallet loading problems.

properties	large objects		small items								pattern restrictions						objectives				
	assortment		shape				assortment		availability		geometry				organization						
			form		orientation				number/figure		dis-tance	separation			number of stages						
													orthogonal								
sources / author(s) and year of publication	variable	fixed	rectangular	non-rectangular	any	90 degree-turns (parallel)	fixed	variable	unlimited	limited	between items	non-orthogonal	nested patterns	guillotine patterns with unlimited stages	single-stage	multi-stage	trim loss minimization (absolutely)	trim loss minimization (relatively)	value maximization	stability	layout optimization
Steudel (1979)		●	●			●	●		●				●		●				●	●	
Smith/Cani de (1980)		●	●			●	●			●			●		●		●				
Bischoff/Dowsland (1982b)		●	●			●	●		●				●		●				●	●	
Dori/Ben-Bassat (1984)		●		●	●		●			●		●			●			●			
Dowsland (1984a)	●		●			●		●		●			●		●				●		
Dowsland (1984b)	●		●			●		●		●			●		●				●		
Steudel (1984)	●		●			●	●			●			●		●				●		
Carpenter/Dowsland (1985)		●	●			●	●			●			●		●				●	●	●
Dowsland (1985a)		●	●			●	●			●			●		●				●		
Dowsland (1985c)		●	●			●	●			●			●		●				●		

Isermann (1985)-1		●	●			●	●			●			●		●				●	●	
Isermann (1985)-2		●	●			●	●			●			●		●				●	●	
Dowsland/Dowsland (1986)		●	●			●	●		●				●		●				●		
Puls/Tanchoco (1986)		●	●			●	●			●	●			●	●				●		
Dowsland (1987a)		●	●			●	●			●			●		●				●		
Dowsland (1987b)		●	●			●	●			●			●		●						●
Isermann (1988)-1		●	●			●	●			●			●		●				●	●	
Isermann (1988)-2	●		●			●		●		●			●		●				●	●	
Daniels/Ghandforoush (1990)		●	●			●	●						●		●				●		
Naujoks (1990)		●	●			●	●			●			●		●				●		

Table 9.2: PL2-problems

9.2 Three-dimensional Pallet Loading Type (PL3)

In the analysed literature, two problems appear concerning the PL3-type, which present the same properties for all of the attributes based on their logical structure. Table 9.3 presents these characteristics. One should note that both of these contributions, Peleg/Peleg (1976) and Han/Knott (1989), deal with theoretical/methodical studies.

9.3 Actual Pallet Loading Problems

Two-dimensional pallet loading types are exclusively represented in application-orientated studies. As mentioned above, the majority of these contributions deal with the problem of loading pallets with delivery boxes, and several applicable software packages are available (Käschel/Mädler/Richter 1991). These programmes particularly reveal the connection between packaging planning and the actual investigation by determining whether measurement variations in delivery lead to better results in packing.

attributes			properties			
dimensionality			1-dimensional	2-dimensional	3-DIMENSIONAL	
type of assignment			TYPE II	Type III	Type IV	
large objects	type of quantity measurement		discrete	continuous		
	figure	form	rectangular	non-rectangular		
	assortment	homogeneous	fixed	variable		
		heterogeneous	fixed	variable		
	availability	number/figure	1	limited	unlimited	mixed
		sequence	no order	partial order	complete order	
small items	type of quantity measurement		discrete	continuous		
	figure	form	rectangular	non-rectangular		
		orientation	fixed	optional	90° turn (parallel)	90° turn (optional)
	assortment	homogeneous	FIXED	variable		
		heterogeneous	fixed	variable	few objects per figure	many objects per figure
	availability	number/figure	limited	unlimited		
		sequence	no order	partial order	complete order	
		dates	identical time references	different time references		
pattern restrictions	geometric characteristics	distancing restrictions	none	between items	to the edges	
		frequency limitations	none	limited combinations of figures	limited number of figures	limited number of items
		separation restrictions	non-orthogonal patterns	nested patterns	guillotine patterns (limited stages)	guillotine patterns (unlimited stages)
	operational characteristics	number of stages	single-stage processes	multi-stage processes		
		pattern sequence	no restrictions	with restrictions		
		pattern frequency	no restrictions	limited number of different patterns	minimum number of identical patterns	
	dynamics of allocation		dynamic	static: off-line	static: on-line	
objectives			input minimization	trim loss minimization	value maximization	others
status of information			deterministic	stochastic		
variability of data			fixed data	variable data		
solution methods			object-orientated methods	single-patterned methods	multi-patterned methods	

Table 9.3 Common characteristics of the PL3-type

10. Conclusions

The significance of the decision-orientated definition of C&P types developed in this book is two-fold: firstly, it supports theoretical insight and, secondly, it offers possibilities for practical use.

On a theoretical level a systematic catalogue of characteristics is available, based on the logical structure of C&P problems. It serves as an instrument to characterize the diverse actual phenomena of cutting and packing so that similarities or differences essential for decision-making are revealed. It can simplify the recognition, analysis and formulation of practical C&P problems so that application guidelines can be formulated for solution procedures to be selected or developed. Furthermore, the catalogue of characteristics presents a consistent terminology for C&P problems.

The hierarchy of types, based on the catalogue presented in the book, groups such properties into clusters and assigns them to types which can sufficiently clarify the essential characteristics for the solution of a group of C&P problems. It thus defines relationships between C&P problems in respect of characteristics particularly relevant to solution approaches. By its hierarchical structure, a system of characteristics is established, which is orientated on the significance that each one has in the solution of C&P problems.

This hierarchy of types is not meant to be complete, as it has been designed according to a limited body of literature analyzed. Nevertheless, it is of general importance when it is seen as a systematic approach which can be expanded upon when necessary. Experiences and investigations up to this point have shown that, for example, the extension of the area of study into abstract C&P problems (i.e. non-spatial dimensions) will result in a catalogue in which what is presented in this study plays a part.

For practical purposes, a systematic overview of sources on C&P problems drawn from literature in various disciplines has been made available. The description of these problems with the aid of their characteristics and their grouping respective to the hierarchy of types, allows for a speedy access to existing knowledge and a direct comparison with those contributions relevant to a particular type. Furthermore, by recording all of this data in a data base, an instrument is also available which offers possibilities for various other literary investigations (cf. Appendix III and footnote 1 on page 6).

It may be emphasized once again that for reasons given in Section 3.1 in several sources it was difficult and at times impossible to assess characteristics. This could perhaps explain some inconsistencies in the matrices of Chapters 6 to 9. It has, however, no significant effect on the fundamental, theoretical findings of this investigation, but can naturally lead to erroneous or distorted illustrations or assignments of individual contributions. As a rule, only the authors themselves can properly correct such distortions.

To our knowledge, the recorded sources are largely complete up to 1990 and are at least representative of the scientific findings up to this point. Although sources appearing after 1990 have not been included in the typology, they are listed in the appendix. The systematic literary assessment is still being pursued and abstracts from current sources are continually being published in the newsletter of the Special Interest Group on Cutting and Packing (SICUP-Bulletin), which appears semi-annually.

It is our hope that with the catalogue of characteristics, the hierarchy of types and the literary overview, a basis of knowledge has in fact been established which can serve as a decisive aid in the selection, development and application of procedures for solving C&P problems. The data base could naturally serve for other investigation purposes as well.

Appendix

I. A Bibliography of Further C&P-Problems

A. Published Surveys

AUTHORS	TYPES										SURVEY ABOUT: Methods Problems Software Theory
	BIN-PACKING (BP)		CUTTING-STOCK (CS)			KNAPSACK (KS)			PALLET-LOADING (PL)		
	1-dim.	2-dim.	1-dim.	2-dim.	3-dim.	1-dim.	2-dim.	3-dim.	2-dim.	3-dim.	
Gilmore/Gomory (1965)			●	●	●						P/M
Falthauser (1969)			●	●							P/M
Höfer (1969)			●	●							P/M
Brown (1971)	●		●	●	●	●	●	●			P/M
Salkin/de Kluyver (1975)						●	●	●			M
Golden (1976)	●		●								P/M
Hinxman (1980)			●	●		(●)					P/M
Garey/Johnson (1981)	●	●									M
Johnston (1981)			●								P
Müller-Merbach (1981)			●	●	●						P
Coffman (1982)	●	●									M
Israni/Sanders (1982)			●	●							M
Lewis/Parker (1982)	(●)					●					M
Rayward-Smith/Shing (1983)	●	●									M
Sarin (1983b)				●							M
Bare et al. (1984)			●	●	●						M
Coffman et al. (1984)	●	●									P/M
Dyckhoff et al. (1984)			●	●	(●)						M
Dowsland (1985b)		●					●		●	●	P/M
Dyckhoff et al. (1985)			●	●	(●)						P
Israni/Sanders (1985)		●									M
Keber (1985)							●	●	●	●	M
Dudzinski/Walukiewicz (1987)						●	●				M
Dyckhoff et al. (1987)			●	●							S
Martello/Toth (1987)						●	●				M
Terno et al. (1987)			●	●							P/M
Dyckhoff (1988a)			●	●	●						T
Dyckhoff (1988b)			●	●	●						T
Dyckhoff et al. (1988a)			●	●	●						P
Dyckhoff et al. (1988b)			●	●							S
Dyckhoff et al. (1988c)			●	●							S
Exeler (1988)									●	●	M
Mannchen (1989)							●	●	●	●	P/M
Dyckhoff (1990)	●	●	●	●	●	●	●	●	●	●	P/M
Martello/Toth (1990)	●					●	●				P/M

Table A.1: Types from Published Surveys (up to 1991)

B. Literary References not Closely Analysed [1]

Those sources on cutting and packing appearing before 1990 but not closely analysed are classified in Table A.2 according to whether they handle abstract, related, mathematical, cutting, packing, branchspecific or other problems. Software is listed separately. Additionally, the table differenciates between published and non-published sources.

Sources concerning abstract problems include problems of assembly line balancing, capital budgeting, layout problems, machine scheduling, coin changing, memory allocation and others. Under the term 'related problems', sources on assortment problems, problems of packaging organization and other related contributions, such as space arrangement problems are listed. 'Mathematical problems' include publications on bin packing, knapsack problems and other sources usually largely concerned with proofs. Under the terms 'cutting problems', 'packing problems' and 'branch-specific problems' are listed unpublished contributions or those published in practitioner magazines. Unpublished surveys, bibliographies and other sources not fitting to the assignment are listed under the term 'others'.

The total number of problems reaches 356, of which 317 are published and 39 unpublished. Of the problems not closely analysed, 74 are considered abstract, 41 related, 129 mathematical, 29 cutting, 29 packing, 36 related to a specific branch of industry, 10 software descriptions, and 8 'others'.

It should be noted that some sources may appear repeatedly and that only the essential parts have been considered for the assignment. It should also be noted again[2] that the synopsis of sources in Table A.2 is by no means complete.

1) Literary References were analysed up to 1991.

2) Cf. footnote 2 on page 17.

		published	unpublished[3]
abstract problems	assembly line balancing	15	0
	capital budgeting	4	0
	layout problem	14	0
	maschine scheduling	10	0
	coin changing	3	0
	memory allocation	12	0
	others	15	1
related problems	assortment problem	18	0
	design	8	0
	others	15	0
mathematical problems	bin packing	16	2
	knapsack problem	103	1
	others	6	1
cutting problems[1]	one-dimensional	6	6
	two-dimensional	7	9
	three-dimensional	1	0
packing problems[1]	one-dimensional	0	0
	two-dimensional	7	2
	three-dimensional	15	5
branchspecific problems[1]	timber industry	6	1
	paper industry	16	1
	steel industry	4	1
	textile industry	7	0
software		5	5
others[2]		4	4

1) unpublished or published in practitioner magazines
2) including unpublished surveys
3) since 1986

Table A.2: Literary References not Closely Analysed

1. Abstract Problems

a. Assembly Line Balancing

published

Salveson (1955)
Jackson (1956)
Bowman (1960)
Held et al. (1963)
Klein (1963)
Gutjahr/Nemhauser (1964)
Garey et al. (1976)
Coffman et al. (1978a)
Pinto et al. (1981)
Wee/Magazine (1982)
Sarker/Shanthikumar (1983)
Agrawal (1985)
Queyranne (1985)
Baybars (1986)
Talbot et al. (1986)

b. Capital Budgeting

published

Lorie/Savage (1955)
Cord (1964)
Woolsey (1972)
Rosenblatt/Sinuany-Stern (1989)

c. Layout Problem

published

Coxeter et al. (1959)
Bruijn de (1969)
Brualdi/Foregger (1974)
Adamowicz/Albano (1976b)
Garfinkel (1977)
George/Robinson (1980)
Chaiken et al. (1981)
Fowler et al. (1981)
Barnes (1982)
Chung et al. (1982a)
Korchemkin (1983)
Schlag et al. (1983)
Stockmeyer (1983)
Kumara et al. (1988)

d. Machine Scheduling

published

Hox (1964)
Struve (1967)
Greenberg (1972)
Simmons (1972)
Paul (1979)
Philipson/Ravindran (1979)
Gehring/Röscher (1988)
Kämpke (1988)
Cheng/Sin (1990)
Stern/Avivi (1990)

e. Coin Changing

published

Tien/Hu (1977)
Martello/Toth (1980a)
Stehling (1983)

f. Memory Allocation

published

Melzak (1966)
Robson (1974)
Chandra/Wong (1975)
Robson (1977)
Coffman/Leung (1979)
Erickson/Luss (1980)
Luss (1983)
Sarin (1983a)
Sarin/Ahn (1983)
Friesen/Langston (1984a)
Langston (1984)
Chen/Liu (1987)

g. Others

published

Erdös/Rogers (1953)
Codd (1960)
Meir/Moser (1968)
Brown (1969)
Chandra et al. (1978)
Brosh (1981)
Zaloom (1982)
Blazewicz/Ecker (1983)
Stadtler (1983)
Coffman/Gilbert (1984)
Cochard/Yost (1985)
Coffman/Gilbert (1985)
Haessler (1985)
Martello/Toth (1985b)
Brockhoff/Braun (1989)

unpublished

Larichev/Furems (1986)

2. Related Problems

a. Assortment Problem

published

Sadowski (1959)
Frank Jr. (1965)
Wilson (1965)
Wolfson (1965)
Cohen (1966)
Eilon/Mallya (1966)
Elmaghraby (1968)
Pentico (1974)
Chambers/Dyson (1976)
Pentico (1976)
Shearn (1976)
Walters (1976a)
Walters (1976b)
Rationalisierungs-Kuratorium der Deutschen Wirtschaft (RKW) (1981)
Diegel/Bocker (1984)
Pentico (1988)
Gochet/Vandebroek (1989)
Gemmill/Sanders (1990)

b. Design

published

Klingst (1965)
Baumgarten/Stolz (1975)
Baumgarten/Gail (1975)
Crankshaw (1979)
Wright (1979)
Jansen/Graefenstein (1988)
author unknown (1988)
Weiser (1988)

c. Others

published

Scheid (1960)
Saaty/Alexander (1975)
Combes (1976)
Roth/Vaughan (1978)
Picard/Queyranne (1981)
Stadtler (1983)
Stoyan (1983)
Tanchoco et al. (1983)
Levcopoulos/Lingas (1984)
Haessler (1985)
Hochbaum/Maass (1985)
Chazelle et al. (1986)
Martin/Stephenson (1988)
Scheithauer/Terno (1988b)
Whitaker/Cammell (1990)

3. Mathematical Problems

a. Bin Packing

published

Kleitman/Krieger (1975)
Kou/Markowsky (1977)
Hoffmann (1981)
Bentley et al. (1984)
Coffman/Gilbert (1984)
Hofri (1984)
Prodinger (1985)
Hochbaum/Shmoys (1986)
Csirik/Galambos (1987)
Krause et al. (1987)
Langston (1987)
Coffman et al. (1988)
Csirik/Totik (1988)
Lueker (1983)
Rhee (1988)
Rhee/Talagrand (1989)

unpublished

Csirik et al. (1990a)
Lai/Chan (1990b)

b. Knapsack Problem

published

Dantzig (1957)
Pandit (1962)
Gilmore/Gomory (1966)
Kolesar (1967)
Shapiro/Wagner (1967)
Weingartner/Ness (1967)
Pierce (1968)
Shapiro (1968)
Garfinkel/Nemhauser (1969)
Greenberg (1969)
Cabot (1970)
Gerhardt (1970)
Greenberg/Hegerich (1970)
Bradley (1971)
Guignard/Spielberg (1972)
Tilanus/Gerhardt (1972)
Faaland (1973)
Ingargiola/Korsh (1973)
Horowitz/Sahni (1974)
Lambe (1974)
Ahrens/Finke (1975)
Fayard/Plateau (1975)
Ibarra/Kim (1975)
Ingargiola/Korsh (1975)
Magazine et al. (1975)
Sahni (1975)
Thesen (1975)
Walukiewicz (1975)
Chandra et al. (1976)
Frieze (1976)
Hirschberg/Wong (1976)
Hu/Lenard (1976)
Kästing (1976)
Nauss (1976)
Fisk (1977)
Ingargiola/Korsh (1977)
Martello/Toth (1977a)
Martello/Toth (1977b)

Balas/Zemel (1978)
Fayard/Plateau (1978)
Hung/Fisk (1978)
Ibaraki et al. (1978)
Lauriere (1978)
Martello/Toth (1978a)
Martello/Toth (1978b)
Müller-Merbach (1978)
Nauss (1978)
Suhl (1978)
Zoltners (1978)
Bulfin et al. (1979)
Denardo/Fox (1979)
Glover/Klingman (1979)
Lawler (1979)
Martello/Toth (1979)
Padberg (1979)
Shih (1979)
Sinha/Zoltners (1979)
Albano/Orsini (1980a)
Balas/Zemel (1980)
Chvatal (1980)
Dembo/Hammer (1980)
Gens/Levner (1980)
Greenberg/Feldman (1980)
Martello/Toth (1980b)
Martello/Toth (1980c)
Padberg (1980)
Toth (1980)
Zemel (1980)
Bitran/Hax (1981)
Faaland (1981)
Johnson/Padberg (1981)
Konno (1981)
Magazine/Oguz (1981)
Martello/Toth (1981a)
Martello/Toth (1981b)
Veliev/Mamedov (1981)
Armstrong et al. (1982)
Fayard/Plateau (1982)
Akinc (1983)
Armstrong et al. (1983)
Brucker (1983)
Dudzinski/Walukiewicz (1984)
Dyer (1984)
Karnin (1984)
Maculan (1984)
Magazine/Oguz (1984)
Martello/Toth (1984)
Zemel (1984)
Aittoniemi/Oehlandt (1985)
Dudzinski/Walukiewicz (1985b)
Greenberg (1985)
Kaufman et al. (1985)
Martello/Toth (1985a)
Orlin (1985)
Plateau/Elkihel (1985)
Greenberg (1986)
Murphy (1986)
Balas et al. (1987)
Marchetti-Spaccamela/Vercellis (1987)
Pirkul (1987)
Yanasse/Soma (1987)
Gottlieb/Rao (1988)
Martello/Toth (1988)

unpublished

Csirik et al. (1990b)

c. Others

published

Smilauer (1962)
Balas (1975)
Akeda/Hori (1976)
Berge (1979)
Marcotte (1985)
Marcotte (1986)

unpublished

Farley (1988a)

4. Cutting Problems

a. One-dimensional

published

Smilauer (1962)
Woolsey (1972)
Reymann (1977)
Haessler (1980b)
Haessler (1985)
Marcotte (1985)

unpublished

Diegel (1988a)
Diegel (1988b)
Diegel (1988c)
Diegel (1988d)
Diegel (1990)
Erol/Kara (1990)

b. Two-dimensional

published

Herz (1972)
Niederhausen (1977a)
Niederhausen (1977b)
Reymann (1977)
Scheithauer/Terno (1985)
Scheithauer/Terno (1986b)
Schlingensiepen (1987)

unpublished

Art (1966)
Hug (1984)
Scheithauer/Terno (1986a)
Scheithauer/Terno (1987)
Farley (1988d)
Scheithauer/Terno (1988a)
Blazewicz et al. (1990b)
Lai/Chan (1990b)
Scheithauer (1990)

c. Three-dimensional

published

Schneider (1979)

5. Packing Problems

a. Two-dimensional

published

Peleg (1971)
Peleg/Orlowski (1971)
Wright (1974)
Akeda/Hori (1975)
Erdös/Graham (1975)
Tanchoco/Agee (1981)
Dagli (1990)

unpublished

Coffman/Shor (1988)
Dowsland/Dowsland (1990)

b. Three-dimensional

published

Ebeling (1971)
author unknown (1973a)
author unknown (1973b)
author unknown (1973c)
author unknown (1973d)
Wright (1973)
Wingerter (1975)
Rockstroh (1978)
Binnenbruck/Schmidt (1979)
Bischoff/Dowsland (1982a)
Bischoff/Dowsland (1983)
Salzer (1983)
Haessler (1985)
Arnold (1988)
Ahbel (1990)

unpublished

Isermann (1987)
Gehring (1988)
Dowsland/Dowsland (1990)
Portmann (1990)
Scheithauer/Terno (1990)

6. Branchspecific Problems

a. Timber Industry

published

Harrell/Smith (1961)
Pnevmaticos/Mann (1972)
Niederhausen/Reuter (1978)
Lembersky/Chi (1986)
Voigt (1987)
Wirsam (1987)

unpublished

Yanasse et al. (1990)

b. Paper Industry

published

Paull/Walter (1955)
Duyne van (1961)
Meerendonk, van den/Schouten (1962)
Hox (1964)
Pegels (1967a)
Struve (1967)
Haessler (1968)
Pierce (1970)
Marconi (1971)
Hartmann (1972)
Kallio (1972)
Bauer (1976)
Haessler (1976)
Köhler (1978)
Goulimis et al. (1986)
Haessler (1988a)

unpublished

MAS-GmbH (1987)

c. Steel Industry

published

Wintgen/Kluge (1961)
Pelzer/Ruth (1966)
Hessling/Richter (1975)
Blank (1976)

unpublished

SCS-GmbH (1986)

d. Textile Industry

published

Wedekind (1968)
Grunwald (1973)
Körner (1974)
Meier (1975)
Zimmermann (1975)
Henne (1977)
Riemer/Hartung (1977)

7. Software

published

Pfefferkorn (1975)
Niederhausen/Reuter (1978)
Reuter (1980)
author unknown (1981)
Dagli (1990)

unpublished

HANIC (1986)
Beyers & Partners (1987)
MAS-GmbH (1987)
VEB Wissenschaftlich-Technisches Zentrum der holzverarbeitenden Industrie (Hrsg.) (1988)
Blazewicz et al. (1990a)

8. Others

published

Pierce (Hrsg.) (1967)
Gardner (1979)
Dyckhoff (1988a)
Dyckhoff (1988b)

unpublished

Dyckhoff/Wäscher (1988)
Sweeney/Ridenour (1989)
Wäscher (1989b)
Lai/Chan (1990a)

C. Most Recent Sources

Anily et al. (1990)
Biro (1990)
Bischoff (1991)
Chauny et al. (1991)
Corominas (1991)
Diegel (1991)
Dowsland (1991a)
Dowsland (1991b)
Dowsland et al. (1991)
Dyckhoff (1991a)
Dyckhoff (1991b)
Exeler (1991)
Farley (1990)
Farley (1991)
Foronda/Carino (1991)
Frerich-Sagurna/Li (1991)
Haberl et al. (1991)
Haessler/Sweeney (1991)
Isermann (1991a)
Isermann (1991b)
Isermann (1991c)
Ivancic et al. (1989)
Johnson (1989)
Käschel et al. (1991)
Kruse (1991)
Luderer/Singer (1991)
Morabito/Arenales (1990)
Morabito/Arenales (1991)
Naujoks (1991)
Ozden (1988)
Pasche (1991)
Schuster (1991)
Tanchoco et al. (1980)
Vasko et al. (1991)
Vel, van de/Shijie (1991)
Walther (1991)
Yanasse et al. (1991)
Yeong/Yue (1991)

II. Brief Description of the Characteristics

DIMENSIONALITY

= minimum number n of relevant dimensions for a geometric description of a pattern

- one-dimensional (n=1)
  e.g. dividing a given length of material into the ordered measurements when objects and items agree according to cross-section and quality
- two-dimensional (n=2)
  e.g. dividing a given sheet into the ordered measurements when objects and items agree according to thickness and quality
- three-dimensional (n=3)
  e.g. dividing a given volume into the ordered measurements when objects and items agree according to quality
- multi-dimensional (n>3)
  e.g. dividing a given volume into the ordered measurements within a certain time interval, when objects and items agree according to quality

TYPE OF ASSIGNMENT

= assignment of large objects to small items

- Type I (all objects, all items)
  = all objects and items are to be assigned
- Type II (all objects, selection of time)
  = a selection of items is to be assigned to all objects
- Type III (selection of objects, all items)
  = all items are to be assigned to a selection of objects
- Type IV (selection of objects, selection of items)
  = selections of both objects and items are to be assigned

CHARACTERISTICS OF OBJECTS AND ITEMS

type of quantity measurement

= type of quantity measurement for objects/items with identical measurements in the relevant dimensions

- discrete
  = frequency and number of objects/items of a certain figure (shape) are measured in integer numbers; e.g. cutting rolls of material with predetermined lengths and widths.
- continuous
  = measuring the total length of an object/item by adding the length of individual pieces without considering the total length to be a relevant dimension, e.g. cutting rolls of material of a given width and total length when the required total length can be cut from several rolls.

figure (shape)

= figure of the object/item is determined by its form, orientation and size

- form
  = contour of the objects/items in two- and multi-dimensional problems
  - rectangular (rectangles or blocks)
  - non-rectangular (e.g. circles, triangles or irregularly-shaped forms)

- orientation
  = position of objects/items in relation to a conceived base
  - fixed
    = objects/items are in a given position; e.g. parallel to the conceived base
  - turns are permitted
    - optional
    - 90-degree-turns parallel to the conceived surface
    - optional 90-degree-turns
    - other predetermined angles

- size
  = measurements of objects/items (e.g. length, width, weight) in the relevant dimensions
  - important aspect: size relations between objects/items
    - relatively small
    - relatively large
    - mixed

assortment

= type and number of existing figures of the objects/items

- homogeneous
  = objects/items have identical measurements (possibly with different orientations; i.e. the figures are congruent, and in case of identical orientation even identical)
  - fixed (measurements of the objects/items are fixed)
  - variable (measurements of the objects/items can vary within a certain scope; e.g. when overloading pallets)
- heterogeneous
  = objects/items have different measurements; i.e. their figures differ in form and/or size
  - fixed (measurements are fixed)
  - variable (measurements of the objects/items can vary within a certain scope; e.g. by variations in cutting the edges in the paper industry)
  - few objects/items per figure
  - many objects/items per figure

availability

= number, sequence and dates of objects/items

- number (per figure)
  = available number
  - limited number (extreme: 1)
  - unlimited number
  - mixed (some objects/items are available in unlimited numbers and some in limited)

- sequence
  - no order
    = no relation between objects/items
  - partial order
    = determining, that e.g. a certain object/item must be packed before an other
  - complete order
    = a definite sequence is given for the objects/items

- dates
  - identical time references
    e.g. identical delivery dates
  - different time references
    e.g. different delivery dates

PATTERN RESTRICTIONS

Geometrical characteristics (pattern restrictions in a narrower sense)
= restrictions in forming cutting or packing patterns

- distancing restrictions
  = a minimal distance between items or the the edge based on physical characteristics
  - between items (= minimal distance between items)
  - to the edges (= minimal distance to the edges)

- frequency limitations
  = type and number of combinations of items and their figures
  - limited combinations of figures
    = per pattern, the combination of the different figures is restricted
  - limited number of figures
    = per pattern, the number of different figures is restricted
  - limited number of items
    = per pattern, the number of items is restricted

- separation restrictions
  = conditions concerning the spatial arrangement of the items in an object
  - non-orthogonal patterns
    = arrangements with any angle to the edges
  - orthogonal patterns
    = arrangements parallel to the edges
  - nested patterns
    = arrangements, which cannot be realized by guillotine cuts
  - guillotine patterns
    = arrangements, which can be realized by guillotine cuts
    - limited stages (number of stages)
      = limitations in the maximum number of guillotine cuts permitted
    - unlimited stages
      = no limitations in the number of guillotine cuts permitted

operational characteristics (organization)
= number of stages, pattern sequence and pattern frequency

- number of stages
  = number of independent work cycles of cutting or packing operations
  - single-stage
    = task is performed in a single, continous operation, with minimal interruptions; only the result of the cutting or packing operation is relevant, i.e. even in processes which actually are carried out by several successive operations, (e.g. in dividing objects with guillotine cuts or in loading containers) are considered to be performed simultaneously in one (timeless) step
  - multi-stage
    = dividing an object within several work cycles; e.g. in the paper industry, when rolls of paper are taken out and set aside until they can be further cut

- pattern sequence
  = processing sequence for the patterns
  - optional
  - restricted
- pattern frequency
  = number of pattern application
  - limited number of different patterns (maximum number of different patterns)
  - minimum number of identical patterns (minimum number of application of identical patterns)

dynamics of allocation

= inclusion of time

- static
  = all objects/items have the same beginning and finishing date
  - On-Line
    = not all objects/items are known in advance
  - Off-Line
    = all objects/items are known in advance
- dynamic
  = objects/items have different beginning and finishing dates
  - with reallocation
    = the spatial arrangement of an item can be rearranged within the object
  - without reallocation
    = the spatial arrangement, once given, cannot be changed

OBJECTIVES

= explicit formulated goals

- input minimization
  = minimizing the number of large objects used
- trim loss minimization (absolute)
  = minimizing trim loss quantity
- trim loss minimization (relative)
  = minimizing trim loss quantity in relation to input quantity
- cost minimization (for use of material or space)
  = minimizing object usage or trim loss according to cost per piece, depending on its figure
- change-over cost minimization
  = minimizing costs for altering cutting or packing facilities
- inventory cost minmization
  = minimizing stock of large objects and small items
- value maximization
  = maximizing the value of packed or cut small items
- others

STATUS OF INFORMATION

= reliability of data

- deterministic (definitive data)
- stochastic (with probabilities)
- uncertain (without probabilities)

VARIABILITY

= quantitative tolerances in delivery of items

- fixed data
  = no quantitative tolerances permitted
- variable data
  = quantitative tolerances allowed within a certain scope

SOLUTION METHODS

= basic structure of the applied solution approach

- object-orientated methods (each item is directly assigned to an object)
- pattern-orientated methods
  = (initially constructing permissable pattern and then assigning the items to the respective 'best' pattern)
  - single-patterned (successive production of one 'best' pattern after the other)
  - multi-patterned (simultaneous production of several 'best' patterns)

PROPERTIES BASED ON REALITY

(evaluation only of application-orientated studies and case studies)

type of objects and items

= naming objects and items

- objects (e.g. glass plates)
- items (e.g. automobile windows)

- branch of industry (e.g. glass industry)

- applied technology (e.g. automatic punching machines)

planning context

= connections to other functions and areas of production planning

- cutting problems
  - isolated planning for cutting
  - with order processing
  - with time scheduling
  - with capacity planning
  - with job shop scheduling
  - with procedure selection
  - with machine control
  - with inventory planning
  - with requirement planning
  - with procurement planning
  - others

- packing problems
  - isolated planning for packing
  - with equilibrium planning
  - with stability planning
  - with packaging planning
  - with transport planning
  - others

software

- <u>practical application</u> (software was developed for practical evaluation)
- <u>simulation</u> (software was developed for purposes of simulation, as in testing algorithms)

III. LARS Data Base System

LARS is an archival, retrieval system which is distributed by WEKA 'Software für Wirtschaft und Verwaltung GmbH' in Frankfurt.

The data base "ZUP" has been developed with the structure presented in Table A.3 for the recording of C&P Literature. The 73 fields of ZUP can be used to construct any masks for the input and output of data and can be combined for purposes of investigation. The mask for data input presented in Figure A.4 was developed for the framework of this investigation as well, and deals with a four-sided mask which is combined here into one unit.

Datenbankdefinition ZUP

Nr.	Kurzb.	Feldname	Anz. Werte	Index	Feldwerte	Z.Lge	Z.An
1	AU	Autor(en)	Mehrfach	mit	Alphanum.	Var.	1
2	TI	Titel	Mehrfach	mit	Alphanum.	Var.	1
3	BIB	Bibliographie	Mehrfach	mit	Alphanum.	Var.	1
4	EJ	Erscheinungsjahr	Einfach	mit	ganze Zahl	4	1
5	SE	Seite	Einfach	ohne	Alphanum.	11	1
6	SW	Stichworte	Mehrfach	mit	Alphanum.	Var.	1
7	KB	Kurzbeschreibung	Mehrfach	ohne	Volltext	Var.	1
8	SO	Standort	Schlüssel	mit	Alphanum.	20	1
9	DT	Dokumententyp	Einfach	mit	Alphanum.	20	1
10	AW	Auswertung	Einfach	mit	0/1	1	1
11	AN	Aufnahme	Einfach	mit	0/1	1	1
12	UBS	Übersicht	Einfach	mit	0/1	1	1
13	AUR	Ausrichtung	Mehrfach	mit	Alphanum.	Var.	1
14	PF	Problemfall	Vektor 1	mit	Alphanum.	Var.	1
15	DI	Dimensionalität	Vektor 1	mit	ganze Zahl	Var.	1
16	GME	Mengenmessung	Vektor 1	mit	Alphanum.	Var.	1
17	GFO	Form (>1 dim)	Vektor 1	mit	Alphanum.	Var.	1
18	GUF	Unterschied. Form	Vektor 1	mit	Alphanum.	Var.	1
19	GGF	gleiche Form	Vektor 1	mit	Alphanum.	Var.	1
20	GAN	Anzahl/Gestalt	Vektor 1	mit	Alphanum.	Var.	1
21	GRF	Reihenfolge	Vektor 1	mit	Alphanum.	Var.	1
22	KME	Mengenmessung	Vektor 1	mit	Alphanum.	Var.	1
23	KFO	Form (>1 dim)	Vektor 1	mit	Alphanum.	Var.	1
24	KOR	Orientierung (>1)	Vektor 1	mit	Alphanum.	Var.	1
25	KGR	Größe	Vektor 1	mit	Alphanum.	Var.	1
26	KUF	unterschied. Form	Vektor 1	mit	Alphanum.	Var.	1
27	KGF	gleiche Form	Vektor 1	mit	Alphanum.	Var.	1
28	KAN	Anzahl/Gestalt	Vektor 1	mit	Alphanum.	Var.	1
29	KRF	Reihenfolge	Vektor 1	mit	Alphanum.	Var.	1
30	KTM	Termine	Vektor 1	mit	Alphanum.	Var.	1
31	MOB	Objekt	Vektor 1	mit	0/1	Var.	1
32	MRA	Rand	Vektor 1	mit	0/1	Var.	1
33	MKB	beschr. Kombinat.	Vektor 1	mit	0/1	Var.	1
34	MGZ	beschr. Gest.zahl	Vektor 1	mit	0/1	Var.	1
35	MOZ	beschr. Obj.zahl	Vektor 1	mit	0/1	Var.	1
36	MTR	Trennfläche (>1d)	Vektor 1	mit	Alphanum.	Var.	1
37	MST	Stufigkeit	Vektor 1	mit	Alphanum.	Var.	1
38	MRF	Reihenfolge	Vektor 1	mit	Alphanum.	Var.	1
39	MUS	Musterhäufigkeit	Vektor 1	mit	Alphanum.	Var.	1
40	MDY	Allok.-Dynamik	Vektor 1	mit	Alphanum.	Var.	1
41	MZU	Zuordnungsart	Vektor 1	mit	Alphanum.	Var.	1
42	INM	Inputminimierung	Vektor 1	mit	0/1	Var.	1
43	VMA	Verschn.min.abs.	Vektor 1	mit	0/1	Var.	1
44	VMR	Verschn.min.rel.	Vektor 1	mit	0/1	Var.	1
45	RKM	Rüstkostenmin.	Vektor 1	mit	0/1	Var.	1
46	KM	Kostenminimierung	Vektor 1	mit	0/1	Var.	1
47	STB	Stabilität	Vektor 1	mit	0/1	Var.	1
48	ZSO	Sonstige	Vektor 1	mit	Alphanum.	Var.	1
49	INF	Informationsstand	Vektor 1	mit	Alphanum.	Var.	1
50	VAR	Variabilität	Vektor 1	mit	Alphanum.	Var.	1
51	BRA	Branche	Vektor 1	mit	Alphanum.	Var.	1
52	GO	Große Objekte	Vektor 1	mit	Alphanum.	Var.	1
53	KL	kleine Objekte	Vektor 1	mit	Alphanum.	Var.	1
54	IZP	isol. Zuschn.plg.	Vektor 1	mit	0/1	Var.	1
55	AAW	Auftragsabwicklg.	Vektor 1	mit	0/1	Var.	1
56	TMP	Terminplanung	Vektor 1	mit	0/1	Var.	1
57	KPL	Kapazitätsplg.	Vektor 1	mit	0/1	Var.	1
58	RPL	Reihenfolgeplg.	Vektor 1	mit	0/1	Var.	1
59	VFP	Verfahrensplg.	Vektor 1	mit	0/1	Var.	1
60	MAS	Maschinensteuerg.	Vektor 1	mit	0/1	Var.	1
61	LBF	Lagerbestandsfüh.	Vektor 1	mit	0/1	Var.	1
62	BPL	Bedarfsplanung	Vektor 1	mit	0/1	Var.	1
63	BSP	Beschaffungsplg.	Vektor 1	mit	0/1	Var.	1
64	PSO	Sonstige	Vektor 1	mit	Alphanum.	Var.	1
65	IPP	isol. Packplg.	Vektor 1	mit	0/1	Var.	1
66	VPP	Verpackungsplg.	Vektor 1	mit	0/1	Var.	1
67	VSO	Sonstige	Vektor 1	mit	Alphanum.	Var.	1
68	SOF	Software	Vektor 1	mit	Alphanum.	Var.	1
69	OBJ	Objektorientiert	Vektor 1	mit	0/1	Var.	1
70	EMU	Ein Muster	Vektor 1	mit	0/1	Var.	1
71	MMU	Mehrere Muster	Vektor 1	mit	0/1	Var.	1
72	WMX	Wertmaximierung	Vektor 1	mit	0/1	Var.	1
73	ABC	abc	Einfach	mit	Alphanum.	1	1

Table A.3: Structure of the C&P data base

ZUP, Dokument 775 Neueintrag

Quellenbezeichnung

Standort Auswertung Aufnahme

Autor(en)

Titel

Bibliographie

Erscheinungsjahr abc

Seite Dokumententyp

Stichworte

Charakteristische Merkmale (1): Übersicht

Ausrichtung

Problemfall↓ Charakteristische Merkmale (2)

Dimensionalität: - dimensional

Eigenschaften der großen Objekte

Mengenmessung Form (>1 dim)

Sortiment: Verfügbarkeit:

Unterschiedliche Form Anzahl/Gestalt

gleiche Form Reihenfolge

Eigenschaften der kleinen Objekte

Mengenmessung Form (>1 dim)

Orientierung (>1) Größe

Sortiment: Verfügbarkeit:

unterschied. Form Anzahl/Gestalt

gleiche Form Reihenfolge

Termine

Problemfall↓ Musterrestriktionen

Geometrische Eigenarten

Abstand: Objekt Häufigkeit: beschränkte Kombination

Rand beschränkte Gestaltzahl

beschränkte Objektzahl

Trennfläche (>1d)

Ablaufbedingte Eigenarten

Stufigkeit Reihenfolge

Musterhäufigkeit

Allokations-Dynamik

Zuordnungsart

Problemfall↓ Besonderheiten der Zuordnung

Zielsetzungen

Inputminimierung Verschnittminimierung (absolut) Rüstkostenminimierung

Kostenminimierung Verschnittminimierung (relativ) Stabilität

Wertmaximierung Sonstige

Informationsstand Variabilität

Realitätsbezogene Merkmale

Branche

Objektart: Große Objekte

Kleine Objekte

Problemfall↓ Planungssituation

Zuschneideprobleme

isolierte Zuschneideplanung: mit Auftragsabwicklung mit Terminplanung

mit Kapazitätsplanung mit Reihenfolgeplanung mit Verfahrensplanung

mit Maschinensteuerung mit Lagerbestandsführung mit Bedarfsplanung

mit Beschaffungsplanung Sonstige

Packprobleme

isolierte Packplanung: mit Verpackungsplanung

Sonstige

Software

Lösungsmethoden Objektorientiert Ein Muster Mehrere Muster

Figure A.4: Registration of the C&P data base

Bibliography

List of Abbreviations for the Journals

ACM	Association for Computing Machinery
AIIE	American Institute of Industrial Engineering Transactions
APF	Ablauf- und Planungsforschung
BuW	Bekleidung und Wäsche
CACM	Communications of the Association for Computing Machinery
CAD	Computer Aided Design
CJ	The Computer Journal
Comp&OR	Computers and Operations Research
ECECSR	Economic Computation and Economic Cybernetic Studies and Research
EJOR	European Journal of Operational Research
IIE	Institute of Industrial Engineering Transactions
IJPD&MM	International Journal of Physical Distribution and Materials Management
IJPR	International Journal of Production Research
IPL	Information Processing Letters
JACM	Journal of the Association for Computing Manufactory
JAlg	Journal of Algorithms
JBA	Journal of Business Administration
JCSS	Journal of Computer and System Sciences
JIE	Journal of Industrial Engineering
JoMES	Journal of Mechanical Engineering Science
JORS	Journal of the Operational Research Society (before:ORQ)
krp	Kostenrechnungspraxis
MP	Mathematical Programming
MS	Management Science
NRLQ	Naval Research Logistics Quarterly
NZOR	New Zealand Operational Research

OR	Operations Research
OR Letters	Operations Research Letters
ORQ	Operations Research Quarterly
ORS	Operations Research - Spektrum
ORV	Operations Research Verfahren (Methods of OR)
PPMCan	Pulp and Paper Magazine of Canada
SIAM Computing	Society of Industrial and Applied Mathematics Journal on Computing
SIAM J.Alg.Dis.Meth.	Society of Industrial and Applied Mathematics Journal on Algebraic and Discrete Methods
WiSt	Wirtschaftswissenschaftliches Studium
ZfB	Zeitschrift für Betriebswirtschaft
ZfbF	Zeitschrift für betriebswirtschaftliche Forschung
ZOR	Zeitschrift für Operations Research
ZwF	Zeitschrift für wirtschaftliche Fertigung und Automation

I. General Literature

Bretzke, W.-R. (1980):
Der Problembezug von Entscheidungsmodellen, Tübingen.

Domschke, W., Drexl, A. (1985):
Logistik: Standorte, München.

Gaitanides, M. (1979):
Vorentscheidungen bei der Formulierung integrierter Produktionsplanungsmodelle, Berlin.

Große-Oetringhaus, W. E. (1974):
Fertigungstypologie unter dem Gesichtspunkt der Fertigungsablaufplanung, Berlin.

Kirsch, W. (1970):
Entscheidungsprozesse, Band 1, Wiesbaden.

Lenz, H. (1987):
Entscheidungsmodell und Entscheidungsrealität - Metatheoretische Überlegungen zum logischen Status von Entscheidungsmodellen und dem Problem ihrer Anwendung auf die Realität, in: Schmidt, R.H., Shor, G. (Hrsg.), Modelle in der Betriebswirtschaftslehre, Wiesbaden, S. 273 - 307.

Müller-Merbach, H. (1977):
Quantitative Entscheidungsvorbereitung - Erwartungen, Enttäuschungen, Chancen, in: DBW 37 (1), S. 11 - 23.

Pfohl, H.-C. (1977):
Problemorientierte Entscheidungsfindung in Organisationen, Berlin/New York.

Schneeweiß, Ch. (1989):
Einführung in die Produktionswirtschaft, dritte, revidierte Auflage, Berlin/Heidelberg/New York/London/Paris/Tokyo.

Wäscher, G. (1984):
Innerbetriebliche Standortplanung - Modelle bei einfacher und mehrfacher Zielsetzung, ZfbF 36, S. 930 - 958.

II. C&P-Literature

Adamowicz, M., Albano, A. (1972):
A Two-Stage Solution of the Cutting-Stock Problem; in: Freiman, C. V. (ed.): Information Processing 71, Amsterdam, pp. 1086 - 1091.

Adamowicz, M., Albano, A. (1976a):
A Solution of the Rectangular Cutting-Stock Problem; in: IEEE Transactions on Systems, Man, and Cybernetics, SMC-6, pp. 302 - 310.

Adamowicz, M., Albano, A. (1976b):
Nesting Two-Dimensional Shapes in Rectangular Modules; in: CAD 8, pp. 27 - 33.

Agrawal, P. K. (1985):
The Related Concept in Assembly Line Balancing; in: IJPR 23, pp. 403 - 421.

Ahbel, T. A. (1990):
Arbeitsvorbereitung an der Rampe; in: Zeitschrift für Logistik 11 (5), pp. 72 - 74.

Ahluwalia, K. G., Saxena, U. (1978):
Development of an Optimal Core Steel Slitting and Inventory Policy; in: AIIE Transactions 10, pp. 399 - 408.

Ahrens, J. H., Finke, G. (1975):
Merging and Sorting Applied to the Zero-One Knapsack Problem; in: OR 23, pp. 1099 - 1109.

Aiello, A., Burattini, E., Massarotti, A., Ventriglia, F. (1980):
"A Posteriori" Evaluation of Bin Packing Approximation Algorithms; in: Discrete Applied Mathematics 2, pp. 159 - 161.

Aittoniemi, L., Oehlandt, K. (1985):
A Note on the Martello-Toth Algorithm for One-Dimensional Knapsack Problems; in: EJOR 20, pp. 117.

Akeda, Y., Hori, M. (1975):
Numerical Test of Palasti's Conjecture on Two-Dimensional Random Packing Density; in: Nature 254, pp. 318 - 319.

Akeda, Y., Hori, M. (1976):
On Random Sequential Packing in Two and Three Dimensions; in: Biometrika 63, pp. 361 - 366.

Akinc, U. (1983):
An Algorithm for the Knapsack Problem; in: IIE Transactions 15, pp. 31 - 36.

Albano, A. (1977):
A Method to Improve Two-Dimensional Layout; in: CAD 9, pp. 48 - 52.

Albano, A., Orsini, R. (1980a):
A Tree Search Approach to the M-Partition and Knapsack Problems; in: CJ 23, pp. 256 - 261.

Albano, A., Orsini, R. (1980b):
A Heuristic Solution of the Rectangular Cutting Stock Problem; in: CJ 23, pp. 338 - 343.

Albano, A., Sapuppo, G. (1980):
Optimal Allocation of Two-Dimensional Irregular Shapes Using Heuristic Search Methods; in: IEEE Transactions on Systems, Man, and Cybernetics SMC-10, pp. 242 - 248.

Albat, R., Wirsam, B. (1989):
Praxisgerechte Zuschnittoptimierungen für Flachglas und Spanplatten; in: Pressmar, D. et al. (eds.), OR Proceedings 1988, Berlin et al., pp. 96 - 102.

Anily, S. et al., Bramel, J., Simchi-Levi, D. (1990):
Worst-Case Analysis of Heuristics for the Bin-Packing Problem with General Cost Structures, unpublished paper, Tel-Aviv.

Armstrong, R. D., Sinha, P., Zoltners, A. A. (1982):
The Multiple-Choice Nested Knapsack Model; in: MS 28, pp. 34 - 43.

Armstrong, R. D., Kung, D. S., Sinha, P., Zoltners, A. A. (1983):
A Computational Study of a Multiple-Choice Knapsack Algorithm; in: ACM Transactions on Mathematical Software 9, pp. 184 - 198.

Arnold, D. (1988):
Lagertechnik - Alles drin - 3D-Nutzung schafft Platz; in: Fabrik 2000, Techno-Tip, pp. 224 - 226.

Art, R. C. (1966):
An Approach to the Two Dimensional, Irregular Cutting Stock Problem; IBM Cambridge Scientific Center Report No. 320 - 2006, Cambridge.

Authors unknown (1973a):
A Novel Approach to Packaging Aids; in: Australian Packaging, February, pp. 20 - 23.

Authors unknown (1973b):
Packaging Aids - Part 2 - Calculating Regular Slotted Container Area; in: Australian Packaging, March, pp. 44 - 47.

Authors unknown (1973c):
Packaging Aids - Part 3 - How to Find the Shape when Your Fibre Container Holds a Number of Individual Packages; in: Australian Packaging, April, pp. 58 - 60.

Authors unknown (1973d):
Packaging Aids - Part 4 - The Corrugated Fibre Container in the Handling and Storage System; in: Australian Packaging, May, pp. 42 - 43.

Authors unknown (1981):
Verschnittoptimierung - Ein unerläßliches Hilfsmittel für den Planer; in: APR 40, pp. 1166 - 1168.

Authors unknown (1988):
Packaging Optimisation - A Case for Savings; in: Packaging News, April, pp. 40 - 41.

Atkins, D. R., Granot, D., Raghavendra, B. G. (1984):
Application of Mathematical Programming to the Plywood Design and Manufacturing Problem; in: MS 30, pp. 1424 - 1441.

Baker, B. S., Coffman, E. G. (1981):
A Tight Asymptotic Bound for Next-Fit-Decreasing Bin-Packing; in: SIAM J.Alg.Dis.Meth. 2, pp. 147 - 152.

Baker, B. S., Schwarz, J. S. (1983):
Shelf Algorithms for Two-Dimensional Packing Problems; in: SIAM Computing 12, pp. 508 - 525.

Baker, B. S., Coffman, E. G., Rivest, R. L. (1980):
Orthogonal Packings in Two Dimensions; in: SIAM Computing 9, pp. 846 - 855.

Baker, B. S., Brown, D. J., Katseff, H. P. (1981):
A 5/4 Algorithm for Two-Dimensional Packing; in: JAlg 24, pp. 348 - 368.

Baker, B. S., Calderbank, A. R., Coffman, E. G., Lagarias, J. C. (1983):
Approximation Algorithms for Maximizing the Number of Squares Packed into a Rectangle; in: SIAM J.Alg.Dis.Meth. 4, pp. 383 - 397.

Balas, E. (1975):
Disjunctive Programming: Cutting Planes from Logical Conditions; in: Mangasarian O.L., Meyer, R. R., Robinson, S. M. (eds.): Nonlinear Programming 2, New York et al., pp. 279 - 312.

Balas, E., Zemel, E. (1978):
Facets of the Knapsack Polytype from Minimal Covers; in: SIAM Journal on Applied Mathematics 34, pp. 119 - 148.

Balas, E., Zemel, E. (1980):
An Algorithm for Large Zero-One Knapsack Problems; in: OR 28, pp. 1130 - 1154.

Balas, E., Nauss, R., Zemel, E. (1987):
Comment on "Some Computational Results on Real 0-1 Knapsack Problems"; in: OR Letters 6, pp. 139 - 140.

Bare, B. B., Briggs, D. G., Roise, J. P., Schreuder, G. F. (1984):
A Survey of Systems Analysis Models in Forestry and the Forest Products Industries; in: EJOR 18, pp. 1 - 18.

Barnes, F. W. (1979):
Packing the Maximum Number of m x n Tiles in a Large p x q Rectangle; in: Discrete Mathematics 26, pp. 93 - 100.

Barnes, F. W. (1982):
Algebraic Theory of Brick Packing I; in: Discrete Mathematics 42, pp. 7 - 26.

Barnett, S., Kynch, G. J. (1967):
Exact Solution of a Simple Cutting Stock Problem; in: OR 15, pp. 1051 - 1056.

Bartholdi, J. J., Vate, J. H. V., Zhang, J. (1989):
Expected Performance of the Shelf Heuristic for 2-Dimensional Packing; in: OR Letters 8, pp. 11 - 16.

Bartmann, D. (1986):
Entwicklung eines DV-Systems zur Auftragsverplanung bei einem Papierhersteller; in: Angewandte Informatik 12, pp. 524 - 532.

Bauer, K. (1976):
Tischrechner optimiert Wellpappenverschnitt - ein Erfahrungsbericht; in: APR 16, pp. 568 - 572.

Baumgarten, H., Gail, M. (1975):
Ladeeinheitenbildung - Zuordnung von Lagergütern und Ladehilfsmitteln; in: Fördern und Heben 25, pp. 29 - 33.

Baumgarten, H., Stolz, G. (1975):
Ladeeinheitenbildung - Verfahren zur Lösung des Stauproblems; in: Fördern und Heben 25, pp. 1443 - 1446.

Baybars, I. (1986):
A Survey of Exact Algorithms for the Simple Assembly Line Balancing Problem; in: MS 32, pp. 909 - 932.

Beasley, J. E. (1985a):
An Exact Two-Dimensional Non-Guillotine Cutting Tree Search Procedure; in: OR 33, pp. 49 - 64.

Beasley, J. E. (1985b):
Bounds for Two-Dimensional Cutting; in: JORS 36, pp. 71 - 74.

Beasley, J. E. (1985c):
An Algorithm for the Two-Dimensional Assortment Problem; in: EJOR 19, pp. 253 - 261.

Beasley, J. E. (1985d):
Algorithms for Unconstrained Two-Dimensional Guillotine Cutting; in: JORS 36, pp. 297 - 306.

Beged-Dov, A. G. (1970):
Some Computational Aspects of the M Paper Mills and P Printers Paper Trim Problem; in: JBA 1 (2), pp. 15 - 34.

Beged-Dov, A. G. (1974):
The Paper Trim Problem: A Variable Demand Analysis; in: AIIE Transactions 6, pp. 84 - 85.

Bengtsson, B. E. (1982):
Packing Rectangular Pieces - A Heuristic Approach; in: CJ 25, pp. 353 - 357.

Bentley, J. L., Johnson, D. S., Leighton, T., McGeoch, C. C. (1983):
An Experimental Study of Bin Packing; in: Proceedings of the 21st Annual Allerton Conference on Communications, Control, and Computing, Urbana, pp. 51 - 60.

Bentley, J. L., Johnson, D. S., Leighton, F. T., McGeoch, C. C., McGeoch, L. A. (1984):
Some Unexpected Expected Behavior Results for Bin Packing; in: Proceedings of the 16th ACM Symposium on Theory of Computing, pp. 219 - 288.

Berge, C. (1979):
Packing Problems and Hypergraph Theory: A Survey; in: Annals of Discrete Mathematics 4, pp. 3 - 37.

Berkey, J. O., Wang, P. Y. (1987):
Two-Dimensional Finite Bin-Packing Algorithms; in: JORS 38, pp. 423 - 429.

Bernhard, R. H. (1967):
Use of Inventories for Reduction of Trim Losses; in: JIE 18, pp. 668 - 670.

Beyers & Partners (1987):
O.M.P. - Optimisation - An Application of the Model Generator - A Cutting Stock Problem, Brasschaat, pamphlet.

Binnenbruck, H. H., Schmidt, E. (1979):
Auswirkungen unterschiedlicher Paletten-Ladehöhen auf die Frachtkosten; in: Transport und Lager 5, pp. 18 - 23.

Biro, M. (1990):
Object-Oriented Interaction in Resource Constrained Scheduling; in: IPL 36, pp. 65 - 67.

Biro, M., Boros, E. (1984):
Network Flows and Non-Guillotine Cutting Patterns; in: EJOR 16, pp. 215 - 221.

Bischoff, E. E. (1991):
Stability Aspects of Pallet Loading; in: ORS 13, pp. 189 - 197.

Bischoff, E. E., Dowsland, W. B. (1982a):
The Desk Top Computer Aids Product Design and Distribution; in: IJPD&MM 12, pp. 12 - 22.

Bischoff, E. E., Dowsland, W. B. (1982b):
An Application of the Micro to Product Design and Distribution; in: JORS 33, pp. 271 - 280.

Bischoff, E. E., Dowsland, W. B. (1983):
Der Mikrocomputer als Hilfsmittel für Produktgestaltung und Vertrieb; in: Neue Verpackung, pp. 232 - 237.

Bischoff, E. E., Marriott, M. D. (1990):
A Comparative Evaluation of Heuristics for Container Loading; in: EJOR 44, pp. 267 - 276.

Bitran, G. R., Hax, A. C. (1981):
Disaggregation and Resource Allocation Using Convex Knapsack Problems with Bounded Variables; in: MS 27, pp. 431 - 441.

Blank, O. (1976):
Verschnittminimierung bei Schneidwerkzeugen mittels EDV automatisieren; in: Maschinenmarkt 82, pp. 1497 - 1500.

Blazewicz, J., Ecker, K. (1983):
A Linear Time Algorithm for Restricted Bin Packing and Scheduling Problems; in: OR Letters 2, pp. 80 - 83.

Blazewicz, J., Drozdowski, M., Soniewicki, B., Walkowiak, R. (1990a):
Decision Support System for Cutting Irregular Shapes - Implementation and Experimental Comparison; Technical University of Poznan, Poland, unpublished paper.

Blazewicz, J., Drozdowski, M., Soniewicki, B., Walkowiak, R. (1990b):
Two-Dimensional Cutting Problem - Basic Complexity Results and Algorithms for Irregular Shapes; Technical University of Poznan, Poland, unpublished paper.

Bleibohm, G., Kuhse, H. D. (1982):
Erfahrungen beim EDV-Einsatz zur zweidimensionalen Verschnittoptimierung; in: Werkstatt und Betrieb 115, pp. 67 - 70.

Bookbinder, J. H., Higginson, J. K. (1986):
Customer Service vs Trim Waste in Corrugated Box Manufacture; in: JORS 37, pp. 1061 - 1071.

Bowman, E. H. (1960):
Assembly-Line Balancing by Linear Programming; in: OR 8, pp. 385 - 389.

Bradley, G. H. (1971):
Transformation of Integer Programs to Knapsack Problems; in: Discrete Mathematics 1, pp. 29 - 45.

Brockhoff, K., Braun, H. (1989):
Optimierung der Sendezeit im Werbefernsehen; in: ZfB 59, pp. 609 - 619.

Brooks, R. L., Smith, C. A. B. , Stone, A. H., Tutte, W. T. (1940):
The Dissection of Rectangles into Squares; in: Duke Mathematical 7, pp. 312 - 340.

Brosh, I. (1981):
Optimal Cargo Allocation on Board a Plane: A Sequential Linear Programming Approach; in: EJOR 8, pp. 40 - 46.

Brown, A. R. (1969):
Selling Television Time: An Optimization Problem; in: CJ 12, pp. 201 - 207.

Brown, A. R. (1971):
Optimum Packing and Depletion: The Computer in Space- and Resource-Usage Problems; London/New York.

Brown, D. J. (1980):
An Improved BL Lower Bound; in: IPL 11, pp. 37 - 39.

Brown, D. J., Baker, B. S., Katseff, H. P. (1982):
Lower Bounds for On-Line Two-Dimensional Packing Algorithms; in: Acta Informatica 18, pp. 207 - 225.

Brualdi, R. A., Foregger, T. H. (1974):
Packing Boxes with Harmonic Bricks; in: Journal of Combinatorical Theory (B) 17, pp. 81 - 114.

Brucker, P. (1983):
An 0 (n)-Algorithm for LP-Knapsacks with a Fixed Number of GUB Constraints; in: ZOR 28, pp. 29 - 40.

Bruijn, N. G. de (1969):
Filling Boxes with Bricks; in: American Mathematics Monthly 76, pp. 702 - 718.

Bruno, J. L., Downey, P. J. (1985):
Probabilistic Bounds for Dual Bin-Packing; in: Acta Informatica 22, pp. 333 - 345.

Bulfin, R. L., Parker, R. G., Shetty, C. M. (1979):
Computational Results with a Branch-and-Bound Algorithm for the General Knapsack Problem; in: NRLQ 26, pp. 41 - 46.

Cabot, A. V. (1970):
An Enumeration Algorithm for Knapsack Problems; in: OR 18, pp. 306 - 311.

Cani, P. de (1978):
A Note on the Two-dimensional Rectangular Cutting-Stock Problem; in: JORS 29, pp. 703 - 706.

Carpenter, H., Dowsland, W. B. (1985):
Practical Considerations of the Pallet-Loading Problem; in: JORS 36, pp. 489 - 497.

Caruso, F. A., Kokat, J. J. (1973):
Coil Slitting - A Shop Floor Solution; in: Industrial Engineering 5, pp. 18 - 23.

Chaiken, S., Kleitman, D. J., Saks, M., Shearer, J. (1981):
Covering Regions by Rectangles; in: SIAM J.Alg.Dis.Meth. 2, pp. 394 - 410.

Chambers, M. L., Dyson, R. G. (1976):
The Cutting Stock Problem in the Flat Glass Industry - Selection of Stock Sizes; in: ORQ 27, pp. 949 - 957.

Chandra, A. K., Wong, C. K. (1975):
Worst-Case Analysis of a Placement Algorithm Related to Storage Allocation; in: SIAM Computing 4, pp. 249 - 263.

Chandra, A. K., Hirschberg, D. S., Wong, C. K. (1976):
Approximate Algorithms for some Generalized Knapsack Problems; in: Theoretical Computer Science 3, pp. 293 - 304.

Chandra, A. K., Hirschberg, D. S., Wong, C. K. (1978):
Bin Packing with Geometric Constraints in Computer Network Design; in: OR 26, pp. 760 - 772.

Chauny, F., Loulou, R., Sadones, S., Soumis, F. (1987):
A Two-Phase Heuristic for Strip Packing: Algorithm and Probabilistic Analysis; in: OR Letters 6, pp. 25 - 33.

Chauny, F., Loulou, R., Sadones, S., Soumis, F. (1991):
A Two-phase Heuristic for the Two-dimensional Cutting-stock Problem; in: JORS 42, pp. 39 - 47.

Chazelle, B. (1983):
The Bottom-Left Bin-Packing Heuristic: An Efficient Implementation; in: IEEE Transactions on Computers C-32, pp. 697 - 707.

Chazelle, B., Drysdale, R. L., Lee, D. T. (1986):
Computing the Largest Empty Rectangle; in: SIAM Computing 15, pp. 300 - 315.

Chen, C. H., Liu, M. L. (1987):
An Application of Bin-Packing to Building Block Placement; in: IEEE International Symposium on Circuits and Systems, pp. 576 - 579.

Cheng, R. M. H., Pila, A. W. (1978):
Maximized Utilization of Surface Areas with Defects by the Dynamic Programming Approach; in: ISA Transactions 17 (4), pp. 61 - 69.

Cheng, T. C. E., Sin, C. C. S. (1990):
A State-of-the-Art Review of Parallel-Machine Scheduling Research; in: EJOR 47, pp. 271 - 292.

Christofides, N., Whitlock, C. (1977):
An Algorithm for Two-Dimensional Cutting Problems; in: OR 25, pp. 30 - 44.

Chung, F. R. K., Gilbert, E. N., Graham, R. L. (1982a):
Tiling Rectangles with Rectangles; in: Mathematics Magazine 55, pp. 286 - 291.

Chung, F. R. K., Garey, M. R., Johnson, D. J. (1982b):
On Packing Two-Dimensional Bins; in: SIAM J.Alg.Dis.Meth. 3, pp. 66 - 76.

Chvatal, V. (1980):
Hard Knapsack Problems; in: OR 28, pp. 1402 - 1411.

Chvatal, V. (1983):
Linear Programming, New York/San Francisco, in particular: Chapter 13, "The Cutting-Stock Problem"; pp. 195 - 212.

Cochard, D. D., Yost, K. A. (1985):
Improving Utilization of Air Force Cargo Aircraft; in: Interfaces 15 (1), pp. 53 - 68.

Codd, E. F. (1960):
Multiprogram Scheduling; in: CACM 3, Part 1-4, pp. 347 - 350, 413 - 418.

Coffield, D. R., Crisp, R. M. (1976):
Extensions to the Coil Slitting Problem; in: IJPR 14, pp. 625 - 630.

Coffman, E. G. (1982):
An Introduction to Proof Techniques for Bin-Packing Approximation Algorithms; in: Dempster, M. A. H. et al. (eds.): Deterministic and Stochastic Scheduling, Amsterdam, pp. 245 - 270.

Coffman, E. G., Gilbert, E. N. (1984):
Dynamic, First-Fit Packings in Two ore More Dimensions; in: Information and Control 61, pp. 1 - 14.

Coffman, E. G., Gilbert, E. N. (1985):
On the Expected Relative Performance of List Scheduling; in: OR 33, pp. 548 - 561.

Coffman, E. G., Lagarias, J. C. (1989):
Algorithms for Packing Squares: A Probabilistic Analysis; in: SIAM Computing 18, pp. 166 - 185.

Coffman, E. G., Leung, J. Y. T. (1979):
Combinatorial Analysis of an Efficient Algorithm for Processor and Storage Allocation; in: SIAM Computing 8, pp. 202 - 217.

Coffman, E. G., Lueker, G. S. (1991):
Probabilsitic Analysis of Packing and Partitioning Algorithms; New York et al.

Coffman, E. G., Shor, P. W. (1988):
Packings in Two Dimensions: Average-Case Analysis of Algorithms; Murray Hill, unpublished paper.

Coffman, E. G., Shor, P. W. (1990):
Average-Case Analysis of Cutting and Packing in Two Dimensions; in: EJOR 44, pp. 134 - 144.

Coffman, E. G., Garey, M. R., Johnson, D. S. (1978a):
An Application of Bin-Packing to Multiprocessor Scheduling; in: SIAM Computing 7, pp. 1 - 17.

Coffman, E. G., Leung, J. Y. T., Ting, D. W. (1978b):
Bin Packing: Maximizing the Number of Pieces Packed; in: Acta Informatica 9, pp. 263 - 271.

Coffman, E. G., Garey, M. R., Johnson, D. S., Tarjan, R. E. (1980a):
Performance Bounds for Level-Oriented Two-Dimensional Packing Algorithms; in: SIAM Computing 9, pp. 808 - 826.

Coffman, E. G., So, K., Hofri, M., Yao, A. C. (1980b):
A Stochastic Model of Bin-Packing; in: Information and Control 44, pp. 105 - 115.

Coffman, E. G., Garey, M. R., Johnson, D. S. (1983):
Dynamic Bin Packing; in: SIAM Computing 12, pp. 227 - 258.

Coffman, E. G., Garey, M. R., Johnson, D. S. (1984):
Approximation Algorithms for Bin-Packing - An Updated Survey; in: Ausiello, G. et al. (eds.): Approximation Algorithms for Computer System Design, Wien 1984, pp. 49 - 106.

Coffman, E. G., Lueker, G. S., Rinnooy Kan, A. H. G. (1988):
Asymptotic Methods in the Probabilistic Analysis of Sequencing and Packing Heuristics; in: MS 34, pp. 266 - 290.

Cohen, G. D. (1966):
Comments on a Paper by M. L. Wolfson: "Selecting the Best Lengths to Stock"; in: OR 14, p. 341.

Combes, L. (1976):
Packing Rectangles into Rectangular Arrangements; in: Environment and Planning B, pp. 3 - 32.

Coppersmith, D., Raghavan, P. (1989):
Multidimensional On-Line Bin Packing: Algorithms and Worst-Case Analysis; in: OR Letters 8, pp. 17 - 20.

Cord, J. (1964):
A Method for Allocating Funds to Investment Projects when Returns are Subject to Uncertainty; in: MS 10, pp. 335 - 341.

Corominas, A. (1991):
Procedures for Solving a 1-Dimensional Cutting Problem; Document Intern de Treball 91/03, Barcelona, unpublished paper.

Coverdale, I., Wharton, F. (1976):
An Improved Heuristic Procedure for a Nonlinear Cutting Stock Problem; in: MS 23, pp. 78 - 86.

Coxeter, H. S. M., Few, L., Rogers, C. A. (1959):
Covering Space with Equal Spheres; in: Mathematika 6, pp. 147 - 157.

Crankshaw, G. (1979):
The Computer as Packaging Design Tool; in: Package Engineering, pp. 39 - 43.

Csirik, J. (1986):
Bin Packing as a Random Walk: A Note on Knödel's Paper; in: OR Letters 5, pp. 161 - 163.

Csirik, J. (1989):
An On-Line Algorithm for Variable-Sized Bin Packing; in: Acta Informatica 26, pp. 697 - 709.

Csirik, J., Galambos, G. (1986):
An O (n) Bin-Packing Algorithm for Uniformly Distributed Data; in: Computing 36, pp. 313 - 319.

Csirik, J., Galambos, G. (1987):
On the Expected Behaviour of the NF Algorithm for a Dual Bin-Packing Problem; in: Acta Cybernetica 8, pp. 5 - 9.

Csirik, J., Imreh, B. (1989):
On the Worst-Case Performance of the NkF Bin-Packing Heuristic; in: Acta Cybernetica 9, pp. 89 - 105.

Csirik, J., Mate, E. (1986):
The Probabilistic Behaviour of the NFD Bin Packing Algorithm; in: Acta Cybernetica 7, pp. 241 - 245.

Csirik, J., Totik, V. (1988):
Online Algorithms for a Dual Version of Bin Packing; in: Discrete Applied Mathematics 21, pp. 163 - 167.

Csirik, J., Galambos, G., Frenk, J. B. G., Frieze, A. M. , Rinooy Kan, A. H. G. (1986):
A Probabilistic Analysis of the Next Fit Decreasing Bin Packing Heuristic; in: OR Letters 5, pp. 233 - 236.

Csirik, J., Frenk, J. B. G., Labbe, M., Zhang, S. (1990a):
Fast Algorithms for Dual Bin Packing, Econometrisches Institut, Erasmus Universität Rotterdam, unpublished paper.

Csirik, J., Frenk, J. B. G., Labbe, M., Zhang, S. (1990b):
Heuristics for the 0-1 Min-Knapsack Problem, Econometrisches Institut, Erasmus Universität Rotterdam, unpublished paper.

Dagli, C. H. (1990):
Knowledge-Based Systems for Cutting Stock Problems; in: EJOR 44, pp. 160 - 166.

Dagli, C. H., Tatoglu, M. Y. (1987):
An Approach to Two-Dimensional Cutting Stock Problems; in: IJPR 25, pp. 175 - 190.

Daniels, J. J., Ghandforoush, P. (1990):
An Improved Algorithm for the Non-Guillotine-Constrained Cutting- Stock Problem; in: JORS 41, pp. 141 - 149.

Dantzig, G. B. (1957):
Discrete-Variable Extremum Problems; in: OR 5, pp. 266 - 277.

Dembo, R. S., Hammer, P. L. (1980):
A Reduction Algorithm for Knapsack Problems; in: ORV 36, pp. 49 - 60.

Denardo, E. V., Fox, B. L. (1979):
Shortest-Route Methods: 2. Group Knapsacks, Expanded Networks, and Branch-and-Bound; in: OR 27, pp. 548 - 566.

Diegel, A. (1988a):
Integer LP Solution for Full-Reel Sheet Cutting Problems; Natal, unpublished paper.

Diegel, A. (1988b):
Integer LP Solution for Large Trim Problems; Natal, unpublished paper.

Diegel, A. (1988c):
Critical Aspects of Integer Trim Problems; Natal, unpublished paper.

Diegel, A. (1988d):
Critical Sizes in Integer Trim Problems; Natal, unpublished paper.

Diegel, A. (1988e):
Cutting Paper in Richards Bay: Dynamic Local or Global Optimization in the Trim Problem; in: Orion 3, pp. 42 - 55.

Diegel, A. (1990):
Absolute Integer Solution for Very Large Trim Problems; Natal, unpublished paper.

Diegel, A. (1991):
Sixes and Sevens with Fixed Loss: Split N into M and M-1, University of Natal, unpublished paper, Durban.

Diegel, A., Bocker, H. J. (1984):
Optimal Dimensions of Virgin Stock in Cutting Glass to Order; in: Dec. Sc. 15, pp. 260 - 274.

Dori, D., Ben-Bassat, M. (1984):
Efficient Nesting of Congruent Convex Figures; in: CACM 27, pp. 228 - 235.

Dowsland, K. A. (1984a):
The Three-Dimensional Pallet Chart: An Analysis of the Factors Affecting the Set of Feasible Layouts for a Class of Two-Dimensional Packing Problems; in: JORS 35, pp. 895 - 905.

Dowsland, K. A. (1985a):
A Graph-Theoretic Approach to a Pallet Loading Problem; in: NZOR 13, pp. 77 - 86.

Dowsland, K. A. (1985c):
Determining an Upper Bound for a Class of Rectangular Packing Problems; in: Comp&OR 12, pp. 201 - 205.

Dowsland, K. A. (1987a):
A Combined Data-Base and Algorithmic Approach to the Pallet-Loading Problem; in: JORS 38, pp. 341 - 345.

Dowsland, K. A. (1987b):
An Exact Algorithm for the Pallet Loading Problem; in: EJOR 31, pp. 78 - 84.

Dowsland, K. A. (1990):
Efficient Automated Pallet Loading; in: EJOR 44, pp. 232 - 238.

Dowsland, K. A. (1991):
Optimising the Palletisation of Cylinders in Cases; in: ORS 13, pp. 204 - 212.

Dowsland, K. A., Dowsland, W. B. (1986):
Packing it May not Be Easy!; in: New Zealand Mathematics Magazine 22, pp. 190 - 195.

Dowsland, K. A., Dowsland, W. B. (1990):
State of the Art Survey - Packing Problems; Swansea, unpublished paper.

Dowsland, K. A., Dowsland, W. B. (1991):
Packing Problems, Invited Review; unpublished paper.

Dowsland, W. B. (1984b):
The Computer as an Aid to Physical Distribution Management; in: EJOR 15, pp. 160 - 168.

Dowsland, W. B. (1985b):
Two and Three Dimensional Packing Problems and Solution Methods; in: NZOR 13, pp. 1 - 18.

Dowsland, W. B. (1991):
Sensitivity Analysis for Pallet Loading; in: ORS 13, pp. 198 - 203.

Dudzinski, K., Walukiewicz, S. (1984):
A Fast Algorithm for the Linear Multiple-Choice Knapsack Problem; in: OR Letters 3, pp. 205 - 209.

Dudzinski, K., Walukiewicz, S. (1985):
On the Multiperiod Binaray Knapsack Problem; in: Methods of Operations Research 49, pp. 223 - 232.

Dudzinski, K., Walukiewicz, S. (1987):
Exact Methods for the Knapsack Problem and its Generalizations; in: EJOR 28, pp. 3 - 21.

Duta, L. D., Fabian, C. S. (1984):
Solving Cutting-Stock Problems through the Monte-Carlo Method; in: ECECSR 19 (3), pp. 35 - 54.

Duyne, R. W. van (1961):
Linear Programming in the Paper Industry; in: TAPPI 44, pp. 189A - 193A.

Dyckhoff, H. (1981):
A New Linear Programming Approach to the Cutting Stock Problem; in: OR 29, pp. 1092 - 1104.

Dyckhoff, H. (1987), Elastische Kuppelproduktion geometrisch definierter Güter; postdoctoral thesis, Hagen.

Dyckhoff, H. (1988a):
Produktionstheoretische Fundierung industrieller Zuschneideprozesse; in: ORS 10, pp. 77 - 96.

Dyckhoff, H. (1988b):
Production Theoretic Foundation of Cutting and Related Processes; in: Fandel, G., Dyckhoff, H., Reese, J. (eds.): Essays on Production Theory and Planning, Berlin et al., chapter 10, pp. 151 - 180.

Dyckhoff, H. (1990):
A Typology of Cutting and Packing Problems; in: EJOR 44, pp. 145 - 159.

Dyckhoff, H. (1991a):
Bridges Between Two Principal Model Formulations for Cutting Stock Processes; in: Fandel, G., Gehring, H. (eds.): Operations Research, Beiträge zur quantitativen Wirtschaftsforschung, Berlin et al., S. 377 - 385.

Dyckhoff, H. (1991b):
Approaches to Cutting and Packing Problems, in: Pridham, M., O'Brien, C. (eds.): Production Research: Approaching the 21st Century, London et al., pp. 46 - 54.

Dyckhoff, H., Gehring, H. (1982):
Vergleich zweier Modelle zur Lösung eines konkreten Verschnitt- und Lagerbestandsplanungsproblems; in: ORS 3, pp. 193 - 198.

Dyckhoff, H., Gehring, H. (1988):
Trim Loss and Inventory Planning in a Small Textile Firm; in: Fandel, G., Dyckhoff, H., Reese, J. (eds.): Essays on Production Theory and Planning, Berlin et al., chapter 11, pp. 181 - 190.

Dyckhoff, H., Wäscher, G. (1988):
Bibliography of Cutting and Packing; unpublished paper, RWTH Aachen.

Dyckhoff, H., Abel, D., Kruse, H. J., Gal, T. (1984):
Klassifizierung realer Verschnittprobleme; in: ZfbF 36, pp. 913 - 929.

Dyckhoff, H., Kruse, H. J., Abel, D., Gal, T. (1985):
Trim Loss and Related Problems; in: OMEGA 13, pp. 59 - 72.

Dyckhoff, H., Kruse, H. J., Milautzki, U. (1987):
Standardsoftware für Zuschneideprobleme; in: ZwF 82, pp. 472 - 477.

Dyckhoff, H., Kruse, H.J., Abel, D., Gal, T. (1988a):
Classification of Real World Trim Loss Problems; in: Fandel, G., Dyckhoff, H., Reese, J. (eds.): Essays on Production Theory and Planning, Berlin et al., chapter 12, pp. 191 - 208.

Dyckhoff, H., Finke, U., Kruse, H. J. (1988b):
Standard Software for Cutting Stock Management; in: Fandel, G., Dyckhoff, H., Reese, J. (eds.): Essays on Production Theory and Planning, Berlin et al., chapter 13, pp. 209 - 221.

Dyckhoff, H., Finke, U., Kruse, H.-J. (1988c):
Empirische Erhebung über Verschnittsoftware; in: ORS 10, pp. 237 - 247.

Dyer, M. E. (1984):
An O (n) Algorithm for the Multiple-Choice Knapsack Linear Program; in: MP 29, pp. 57 - 63.

Dyson, R. G., Gregory, A. S. (1974):
The Cutting Stock Problem in the Flat Glass Industry; in: ORQ 25, pp. 41 - 53.

Ebeling, C. W. (1971):
Pallet Pattern by Computer; in: Modern Packaging, pp. 66 - 68.

Eilon, S. (1960):
Optimizing the Shearing of Steel Bars; in: JoMES 2, pp. 129 - 142.

Eilon, S. (1984):
An Enhanced Algorithm for the Loading Problem; in: OMEGA 12, pp. 189 - 190.

Eilon, S., Christofides, N. (1971):
The Loading Problem; in: MS 17, pp. 259 - 268.

Eilon, S., Christofides, N. (1972):
On the Loading Problem - A Rejoinder; in: MS 18, pp. 432 - 433.

Eilon, S., Mallya, R. V. (1966):
A Note on the Optimal Division of a Container into Two Compartments; in: IJPR 5, pp. 163 - 169.

Eisemann, K. (1957):
The Trim Problem; in: MS 3, pp. 279 - 284.

Ellinger, T., Asmussen, R., Schirmer, A. (1980):
Materialeinsparung durch optimale Zuschnittplanung; Eschborn.

Ellinger, T., Asmussen, R., Schirmer, A. (1981):
Rationalisierung durch Verschnittoptimierung; Eschborn.

Elmaghraby, S. E. (1968):
A Loading Problem in Process Type Production; in: OR 16, pp. 902 - 914.

Eng, G., Daellenbach, H. G. (1985):
Forest Outturn Optimization by Dantzig-Wolfe Decomposition and Dynamic Programming Column Generation; in: OR 33, pp. 459 - 464.

Erdös, P., Graham, R. L. (1975):
On Packing Squares with Equal Squares; in: Journal of Combinatorial Theory 19, pp. 119 - 123.

Erdös, P., Rogers, C. A. (1953):
The Covering of n-Dimensional Space by Spheres; in: The Journal of the London Mathematical Society 28, pp. 287 - 293.

Erickson, R. E., Luss, H. (1980):
Optimal Sizing of Records used to Store Messages of Various Lengths; in: MS 26, pp. 796 - 809.

Erol, D., Kara, I. (1990):
A Heuristic for Two Dimensional Guillotine Cutting Stock Problem; unpublished paper, Anadolu University.

Eversheim, W., Hemgesberg, G. (1977):
Entwicklung eines Systems zur optimalen zweidimensionalen Verschnittoptimierung in der Einzel- und Kleinserienfertigung mit Hilfe dialogfähiger Rechenanlagen; Forschungsbericht des Landes NRW 2663, Opladen.

Exeler, H. (1988):
Das homogene Packproblem in der betriebswirtschaftlichen Logistik; Heidelberg.

Exeler, H. (1991):
Upper Bounds for the Homogeneous Case of a Two-Dimensional Packing Problem; in: ZOR 35, pp. 45 - 58.

Faaland, B. (1973):
Solution of the Value-Independent Knapsack Problem by Partioning; in: OR 21, pp. 332 - 337.

Faaland, B. H. (1981):
The Multiperiod Knapsack Problem; in: OR 29, pp. 612 - 616.

Faaland, B., Briggs, D. (1984):
Log Bucking and Lumber Manufactoring Using Dynamic Programming; in: MS 30, pp. 245 - 252.

Fabian, C. (1984):
Greedy Algorithm for a Type of Integer Programming Problems; in: Proceedings of the Colloquium on Approximation and Optimization, Claj-Napoen, 25.-27. Oktober, pp. 245 - 254.

Fabian, C., Duta, D. L. (1982):
The Cutting Stock Problem with Several Objective Functions and in Fuzzy Conditions; in: ECECSR 17, pp. 43 - 47.

Faltlhauser, R. (1968):
Verschnittoptimierung; in: Bussmann, K. F., Mertens, P. (eds.): OR und DV bei der Produktionsplanung, Stuttgart, pp. 187 - 197.

Fang, S. C., Lamendola, C. N. (1982):
Optimal Scheduling of the Coil Slitting Problem; in: AIIE Proceedings , pp. 476 - 480.

Farley, A. A. (1983a):
Trim-Loss Pattern Rearrangement and its Relevance to the Flat-Glass Industry; in: EJOR 14, pp. 386 - 392.

Farley, A. A. (1983b):
A Note on Modifying a Two-Dimensional Trim-Loss Algorithm to Deal with Cutting Restrictions; in: EJOR 14, pp. 393 - 395.

Farley, A. A. (1988a):
A Note on Bounding a Class of Linear Programming Problems, Including Cutting Stock Problems; unpublished paper, Melbourne.

Farley, A. A. (1988b):
Mathematical Programming Models for Cutting-Stock Problems in the Clothing Industry; in: JORS 39, pp. 41 - 53.

Farley, A. A. (1988c):
Practical Adaptations of the Gilmore-Gomory Approach to Cutting Stock Problems; in: ORS 10, pp. 113 - 123.

Farley, A. A. (1988d):
Limiting the Number of Each Piece in Two-Dimensional Cutting Stock Patterns; unpublished paper, Melbourne.

Farley, A. A. (1990a):
The Cutting Stock Problem in the Canvas Industry; in: EJOR 44, pp. 247 - 255.

Farley, A. A. (1990b):
Selection of Stockplate Characteristics and Cutting Style for Two Dimensional Cutting Stock Situations; in: EJOR 44, pp. 239 - 246.

Farley, A. A. (1990c):
A Note on Bounding a Class of Linear Programming Problems, Including Cutting Stock Problems; in: OR 38, pp. 922 - 923.

Farley, A. A. (1991):
Planning the Cutting of Photographic Color Paper Rolls for Kodak (Australasia) Pty. Ltd.; in: Interfaces 21 (1), pp. 92 - 106.

Farley, A. A., Richardson, K. V. (1984):
Fixed Charge Problems with Identical Fixed Charges; in: EJOR 18, pp. 245 - 249.

Fayard, D., Plateau, G. (1975):
Resolution of the 0-1 Knapsack Problem: Conmparison of Methods; in: MP 8, pp. 272 - 307.

Fayard, D., Plateau, G. (1978):
On "An Efficient Algorithm for the 0-1 Knapsack Problem, by Robert M. Nauss"; in: MS 24, pp. 918 - 919.

Fayard, D., Plateau, G. (1982):
An Algorithm for the Solution of the 0-1 Knapsack Problem; in: Computing 28, pp. 269 - 287.

Fernandez de la Vega, W., Lueker, G. S. (1981):
Bin Packing Can be Solved within 1+epsilon in Linear Time; in: Combinatoria 1, pp. 349 - 355.

Ferreira, J. S., Neves, M. A., Fonseca e Castro, P. (1990):
A Two-Phase Roll Cutting Problem; in: EJOR 44, pp. 185 - 196.

Filmer, P. J. (1970):
A Sheet Scheduling Model for a Digital Computer; in: Appita 24, pp. 189 - 195.

Fisk, J. C. (1977):
An Initial Bounding Procedure for Use with 0-1 Single Knapsack Algorithms; in: Opsearch 14, pp. 88 - 98.

Fisk, J. C., Hung, M. S. (1979):
A Heuristic Routine for Solving Large Loading Problems; in: NRLQ 26, pp. 643 - 650.

Förstner, K. (1959):
Zur Lösung von Entscheidungsaufgaben bei der Papierherstellung; in: ZfB 29, pp. 693 - 703, 756 - 765.

Förstner, K. (1963):
Die Aufstellung von Produktionsplänen bei der Papiererzeugung; in: ORV 1, pp. 73 - 92.

Foronda, S. U., Carino, H. F. (1991):
A Heuristic Approach to the Lumber Allocation Problem in Hardwood Dimension and Furniture Manufacturing, in: EJOR 54, pp. 151 - 162.

Foster, D. P., Vohra, R. V. (1989):
Probabilistic Analysis of a Heuristic for the Dual Bin Packing Problem; in: IPL 31, pp. 287 - 290.

Fowler, R. J., Paterson, M. S., Tanimoto, L. (1981):
Optimal Packing and Covering in the Plane are NP-Complete; in: IPL 12, pp. 133 - 137.

Frank Jr., C. R. (1965):
A Note on the Assortment Problem; in: MS 11, pp. 724 - 726.

Frederickson, G. N. (1980):
Probabilistic Analysis for Simple One- and Two-Dimensional Bin Packing Algorithms; in: IPL 11, pp. 156 - 161.

Frenk, J. B. G., Galambos, G. (1987):
Hybrid Next-Fit Algorithm for the Two-Dimensional Rectangle Bin-Packing Problem; in: Computing 39, pp. 201 - 217.

Frerich-Sagurna, R., Li, H. (1991):
Packmustergenerierung - Projektbericht und Perspektiven; in: ORS 13, pp. 249 - 253.

Friesen, D. K., Langston, M. A. (1984):
A Storage-Size Selection Problem; in: IPL 18, pp. 295 - 296.

Friesen, D. K., Langston, M. A. (1986):
Variable Sized Bin Packing; in: SIAM Computing 15, pp. 222 - 230.

Frieze, A. M. (1976):
Shortest Path Algorithms for Knapsack Type Problems; in: MP 11, pp. 150 - 157.

Galambos, G. (1986):
Parametric Lower Bound for On-Line Bin-Packing; in: SIAM J.Alg.Dis.Meth. 7, pp. 362 - 367.

Gardner, M. (1979):
Some Packing Problems That Cannot Be Solved by Sitting on The Suitcase; in: Scientific American 241 (4), pp. 22 - 26.

Garey, M. R., Johnson, D. S. (1981):
Approximation Algorithms for Bin Packing Problems: A Survey; in: Ausiello, D., Lucertini, M. (eds.): Analysis and Design of Algorithms in Combinatorial Optimization, CISM Courses and Lectures 266, Wien, pp. 147 - 172.

Garey, M. R., Graham, R. L., Ullman, J. D. (1973):
An Analysis of Some Packing Algorithms; in: Rustin, R. (ed.): Combinatorial Algorithms (Algorithmic Press), New York, pp. 39 - 47.

Garey, M. R., Graham, R. L., Johnson, D. S., Yao, A. C. C. (1976):
Resource Constrained Scheduling as Generalized Bin Packing; in: Journal of Combinatorial Theory (A) 21, pp. 257 - 298.

Garey, M. R., Graham, R. L., Johnson, D. S. (1978):
On a Number-Theoretic Bin Packing Conjecture; in: Proceedings 5th Combinatorics-Colloquium, Amsterdam 1976, pp. 377 - 392.

Garfinkel, R. S. (1977):
Minimizing Wallpaper Waste, Part 1: A Class of Traveling Salesman Problems; in: OR 25, pp. 741 - 751.

Garfinkel, R. S., Nemhauser, G. L. (1969):
The Set-Partitioning Problem: Set Covering with Equality Constraints; in: OR 17, pp. 848 - 856.

Gaudioso, M. (1979):
A Heuristic Approach to the Cutting Stock Problem; in: ORV 32, pp. 113 - 118.

Geerts, J. M. P. (1984):
Mathematical Solution for Optimising the Sawing Pattern of a Log Given its Dimensions and its Defect Core; in: New Zealand Journal of Forestry Science 14, pp. 124 - 134.

Gehring, H. (1988):
Computergestützte Stauplanung von Standardcontainern im Sammelladungsverkehr; unpublished paper, Aachen.

Gehring, H., Röscher, P. (1988):
Heuristische Maschinenbelegungsplanung in der Glasbehälterproduktion; in: ZwF 83, pp. 589 - 593.

Gehring, H., Gal, T., Rödder, W. (1979):
Produktions- und Lagerbestandsplanung mit einem mehrstufigen Produktionsmodell; in: Gaede, K. W. et al. (eds.): Proceedings in Operations Research, Würzburg, S. 390 - 396.

Gehring, H., Menschner, K., Meyer, M. (1990):
A Computer-Based Heuristic for Packing Pooled Shipment Containers; in: EJOR 44, pp. 277 - 288.

Gemmill, D. D., Sanders, J. L. (1990):
Approximate Solutions for the Cutting Stock "Portfolio" Problem; in: EJOR 44, pp. 167 - 174.

Gens, G. V., Levner, E. V. (1980):
Fast Approximation Algorithms for Knapsack Type Problems; in: Optimization Techniques - Part 2, Berlin et al., pp. 185 - 194.

George, J. A., Robinson, D. F. (1980):
A Heuristic for Packing Boxes into a Container; in: Comp&OR 7, pp. 147 - 156.

Gerhardt, C. (1970):
Gedanken zur Lösung des Knapsack-Problems; in: APF 11 (2), pp. 69 - 83.

Gerhardt, C. (1973):
Anmerkungen zum Aufsatz "Eine Näherungslösung für ein spezielles zweidimensionales Verschnittproblem" von W. Heim; in: ZOR 17, pp. B67 - B68.

Gilmore, P. C. (1966):
The Cutting Stock Problem; in: IBM Proceedings on Combinatorial Problems, pp. 211 - 224.

Gilmore, P. C. (1979):
Cutting Stock, Linear Programming, Knapsacking, Dynamic Programming and Integer Programming, Some Interconnections; in: Annals of Discrete Mathematics 4, pp. 217 - 235.

Gilmore, P. C., Gomory, R. E. (1961):
A Linear Programming Approach to the Cutting-Stock Problem Part I; in: OR 9, pp. 849 - 859.

Gilmore, P. C., Gomory, R. E. (1963):
A Linear Programming Approach to the Cutting Stock Problem - Part II; in: OR 11, pp. 863 - 888.

Gilmore, P. C., Gomory, R. E. (1965):
Multistage Cutting Stock Problems of Two and More Dimensions; in: OR 13, pp. 94 - 120.

Gilmore, P. C., Gomory, R. E. (1966):
The Theory and Computation of Knapsack Functions; in: OR 14, pp. 1045 - 1074.

Glover, F., Klingman, D. (1979):
A 0 (n log n) Algorithm for LP Knapsacks with GUB Constraints; in: MP 17, pp. 345 - 361.

Gochet, W., Vandebroek, M. (1989):
A Dynamic Programming Based Heuristic for Industrial Buying of Cardboard; in: EJOR 38, pp. 104 - 112.

Golan, I. (1981):
Performance Bounds for Orthogonal Oriented Two-Dimensional Packing Algorithms; in: SIAM Computing 10, pp. 571 - 582.

Golden, B. L. (1976):
Approaches to the Cutting Stock Problem; in: AIIE Transactions 8, pp. 265 - 274.

Goswami, P. (1973):
Wertanalysebeispiel: Reduktion des Blechverschnitts im Trafobau; in: Industrielle Organisation 42, pp. 343 - 347.

Gottlieb, E. S., Rao, M. R. (1988):
Facets of the Knapsack Polytope Derived From Disjoint and Overlapping Index Configurations; in: OR Letters 7, pp. 95 - 100.

Goulimis, C. (1990):
Optimal Solutions for the Cutting Stock Problem; in: EJOR 44, pp. 197 - 208.

Goulimis, C. N., Bryant, G. F., Poon, M. C. K. (1986):
Deckling in the Paper and Board Industry - A Breakthrough in the Cutting Stock Problem; in: IFAC Instrumentation and Automation in the Paper, Plastics and Polymerization Industries, Ohio, pp. 141 - 146.

Greenberg, H. (1969):
An Algorithm for the Computation of Knapsack Functions; in: Journal of Mathematical Analysis and Applications 26, pp. 159 - 162.

Greenberg, H. (1985):
An Algorithm for the Periodic Solutions in the Knapsack Problem; in: Journal of Mathematical Analysis and Applications 111, pp. 327 - 331.

Greenberg, H. (1986):
On Equivalent Knapsack Problems; in: Discrete Applied Mathematics 14, pp. 263 - 268.

Greenberg, H., Feldman, I. (1980):
A Better Step-Off Algorithm for the Knapsack Problem; in: Discrete Applied Mathematics 2, pp. 21 - 25.

Greenberg, H., Hegerich, R. L. (1970):
A Branch Search Algorithm for the Knapsack Problem; in: MS 16, pp. 327 - 332.

Greenberg, I. (1972):
Application of the Loading Algorithm to Balance Workloads; in: AIIE Transactions 4, pp. 337 - 339.

Gribov, A. B. (1973):
Algorithm for Solving the Problem of a Plane Cutting Layout; in: Cybernetics 9, pp. 1036 - 1043.

Grunwald, R. (1973):
Anregungen zur Rationalisierung im Zuschnitt; in: BuW 13, pp. 1043 - 1046.

Guignard, M. M., Spielberg, K. (1972):
Mixed-Integer Algorithms for the (0,1) Knapsack Problem; in: IBM Journal of Research and Development 16, pp. 424 - 430.

Gutjahr, A. L., Nemhauser, G. L. (1964):
An Algorithm for the Line Balancing Problem; in: MS 11, pp. 308 - 315.

Haberl, J., Nowak, Chr., Stettner, H., Stoiser, G., Woschitz, H. (1991):
A Branch-and-Bound Algorithm for Solving a Fixed Charge Problem in the Profit Optimization of Sawn Timber Production; in: ZOR 35, pp. 151 - 166.

Haessler, R. W. (1968):
An Application of Heuristic Programming to a Nonlinear Cutting Stock Problem Occurring in the Paper Industry, Michigan, doctoral thesis.

Haessler, R. W. (1971):
A Heuristic Programming Solution to a Nonlinear Cutting Stock Problem; in: MS 17, pp. 793 - 802.

Haessler, R. W. (1975):
Controlling Cutting Pattern Changes in One-Dimensional Trim Problems; in: OR 23, pp. 483 - 493.

Haessler, R. W. (1976):
Single-Machine Roll Trim Problems and Solution Procedures; in: TAPPI 59, pp. 145 - 149.

Haessler, R. W. (1978):
A Procedure for Solving the 1.5-Dimensional Coil Slitting Problem; in: AIIE Transactions 10, pp. 70 - 75.

Haessler, R. W. (1979):
Solving the Two-Stage Cutting Stock Problem; in: OMEGA 7, pp. 145 - 151.

Haessler, R. W. (1980a):
A Note on Computational Modifications to the Gilmore-Gomory Cutting Stock Algorithm; in: OR 28, pp. 1001 - 1005.

Haessler, R. W. (1980b):
Multimachine Roll Trim - Problems and Solutions; in: TAPPI 63, pp. 71 - 74.

Haessler, R. W. (1983):
Developing an Industrial-Grade Heuristic Problem-Solving Procedure; in: Interfaces 13 (3), pp. 62 - 71.

Haessler, R. W. (1985):
Production Planning and Scheduling for an Integrated Container Company; in: Automatica 21, pp. 445 - 452.

Haessler, R. W. (1988a):
A New Generation of Paper Machine Trim Programs; in: TAPPI Journal, August, pp. 127 - 130.

Haessler, R. W. (1988b):
Selection and Design of Heuristic Procedures for Solving Roll Trim Problems; in: MS 34, pp. 1460 - 1471.

Haessler, R. W., Sweeney, P. E. (1991):
Cutting Stock Problems and Solution Procedures; in: EJOR 54, pp. 141 - 150.

Haessler, R. W., Talbot, F. B. (1983):
A 0-1 Model for Solving the Corrugator Trim Problem; in: MS 29, pp. 200 - 209.

Haessler, R. W., Talbot, F. B. (1990):
Load Planning for Shipments of Low Density Products; in: EJOR 44, pp. 289 - 299.

Haessler, R. W., Vonderembse, M. A. (1979):
A Procedure for Solving the Master Slab Cutting Stock Problem in the Steel Industry; in: AIIE Transactions 11, pp. 160 - 165.

Hahn, S. G. (1968):
On the Optimal Cutting of Defective Sheets; in: OR 16, pp. 1100 - 1114.

Haims, M. J., Freeman, H. (1970):
A Multistage Solution of the Template-Layout Problem; in: IEEE Transactions on Systems Science and Cybernetics SSC-6, pp. 145 - 151.

Hall, N. G. (1988):
Bin Packing Problems in One Dimension: Heuristic Solutions and Confidence Intervals; in: Comp&OR 15, pp. 171 - 177.

Han, C. P., Knott, K., Egbelu, P. J. (1989):
A Heuristic Approach to the Three-Dimensional Cargo-Loading Problem; in: IJPR 27, pp. 757 - 774.

HANIC (1986):
OPTIPLAN - Die Verschnittoptimierung für glasverarbeitende Betriebe, pamphlet.

Hardley, C. J. (1976):
Optimal Cutting of Zinc-Coated Steel Strip; in: NZOR 4, pp. 92 - 100.

Harrell, C., Smith, G. W. (1961):
Linear Programming in Log Production; in: Forest Products Journal 11, pp. 8 - 11.

Harrison, P. (1983):
A Multi-Objective Decision Problem: The Furniture Manufacturer's 2-Dimensional Cutting or Trim Problem; in: French, S. et al. (eds.): Multi-Objective Decision Making, London et al., pp. 231 - 236.

Harrison, P. (1984):
The Two Dimensional Cutting Problem of the Furniture Manufacturer; in: Brans, J. P. (ed.): Operational Research '84, Amsterdam, pp. 625 - 640.

Hartmann, W. (1972):
Schneidplanung im Dialog mit dem Rechner - Minimierung der Verschnittverluste in der Papierindustrie; in: Data Report 7, pp. 36 - 39.

Heicken, K., König, W. (1980):
Integration eines heuristisch-optimierenden Verfahrens zur Lösung eines eindimensionalen Verschnittproblems in einem EDV-gestützten Produktionsplanungs- und -steuerungssystem; in: ORS 1, pp. 251 - 259.

Heim, W. (1971):
Eine Näherungslösung für ein spezielles zweidimensionales Verschnittproblem; in: APF 12, pp. 42 - 47.

Held, M., Karp, R. M., Shareshian, R. (1963):
Assembly-Line Balancing - Dynamic Programming with Precedence Constraints; in: OR 11, pp. 442 - 459.

Henne, H. (1977):
Automation im Zuschnitt; in: BuW 5, pp. 255 - 261.

Herz, J. C. (1972):
Recursive Computational Procedure for Two-Dimensional Stock Cutting; in: IBM Journal of Research and Development 16, pp. 462 - 469.

Hessling, R., Richter, H. (1975):
Auftragsabwicklung für ein Blechwalzwerk; in: IBM Nachrichten 25 (224), pp. 49 - 55.

Hinxman, A. I. (1977):
A Two-Dimensional Trim-Loss Problem with Sequencing Constraints; in: Advance Papers of IJCAI-77, M.I.T., pp. 859 - 864.

Hinxman, A. I. (1980):
The Trim-Loss and Assortment Problems: A Survey; in: EJOR 5, pp. 8 - 18.

Hirschberg, D. S., Wong, C. K. (1976):
A Polynomial-Time Algorithm for the Knapsack Problem with Two Variables; in: JACM 23, pp. 147 - 154.

Hochbaum, D. S., Maass, W. (1985):
Approximation Schemes for Covering and Packing Problems in Image Processing and VLSI; in: ACM 32, pp. 130 - 136.

Hochbaum, D. S., Shmoys, D. B. (1986):
A Packing Problem You can Almost Solve by Sitting on Your Suitcase; in: SIAM J.Alg.Dis.Meth. 7, pp. 247 - 257.

Hodgson, T. J. (1982):
A Combined Approach to the Pallet Loading Problem; in: IIE Transactions 14, pp. 175 - 182.

Hodgson, T. J., Hughes, D. S., Martin-Vega, L. A. (1983):
A Note on a Combined Approach to the Pallet Loading Problem; in: IIE Transactions 15, pp. 268 - 271.

Höfer, G. (1969):
Zuschnittprobleme; in: Fischer, H. (ed.): Operationsforschung in der sozialistischen Wirtschaft; Berlin, chapter 8, pp. 301 - 322.

Hoffmann, U. (1981):
Stochastische Packungsalgorithmen; Stuttgart, doctoral thesis.

Hoffmann, U. (1982):
A Class of Simple Stochastic Online Bin Packing Algorithms; in: Computing 29, pp. 227 - 239.

Hofri, M. (1980):
Two-Dimensional Packing: Expected Performance of Simple Level Algorithms; in: Information and Control 45, pp. 1 - 17.

Hofri, M. (1984):
A Probabilistic Analysis of the Next-Fit Bin Packing Algorithm; in: JAlg 5, pp. 547 - 556.

Horowitz, E., Sahni, S. (1974):
Computing Partitions with Applications to the Knapsack Problem; in: JACM 21, pp. 277 - 292.

Hox, G. (1964):
Erweiterte Anwendung von Linear Programming bei der Belegung von Papiermaschinen; in: APF 5, pp. 26 - 33.

Hu, T. C., Lenard, M. L. (1976):
Optimality of a Heuristic Solution for a Class of Knapsack Problems; in: OR 24, pp. 193 - 196.

Hug, R. (1984):
Dispositiver Dialog in der Produktionsplanung am Beispiel eines zweidimensionalen Guillotineschnittproblems; Zürich, doctoral thesis.

Hummler, B., (1989):
Technologie im Anzug - Bekleidungsindustrie jetzt wettbewerbsfähig?; in: Computer Magazin, pp. 21 - 24.

Hung, M. S., Brown, J. R. (1978):
An Algorithm for a Class of Loading Problems; in: NRLQ 25, pp. 289 - 297.

Hung, M. S., Fisk, J. C. (1978):
An Algorithm for 0-1 Multiple-Knapsack Problems; in: NRLQ 25, pp. 571 - 579.

Ibaraki, T., Hasegawa, T., Teranaka, K., Iwase, J. (1978):
The Multiple-Choice Knapsack Problem; in: JORS of Japan 21, pp. 59 - 95.

Ibarra, O. H., Kim, C. E. (1975):
Fast Approximation Algorithms for the Knapsack and Sum of Subset Problems; in: JACM 22, pp. 463 - 468.

Ingargiola, G. P., Korsh, J. F. (1973):
Reduction Algorithm for Zero-One Single Knapsack Problems; in: MS 20, pp. 460 - 463.

Ingargiola, G. P., Korsh, J. F. (1975):
An Algorithm for the Solution of 0-1 Loading Problems; in: OR 23, pp. 1110 - 1119.

Ingargiola, G. P., Korsh, J. F. (1977):
A General Algorithm for One-Dimensional Knapsack Problems; in: OR 25, pp. 752 - 759.

Isermann, H. (1985):
Stapelung von rechteckigen Versandgebinden auf Paletten - Generierung von Stapelplänen im Dialog mit einem PC; in: Ohse, D. et al. (eds.): OR Proceedings 1984, Berlin et al., pp. 354 - 362.

Isermann, H. (1987):
Ein Planungssystem zur Optimierung der Palettenbildung mit kongruenten rechteckigen Versandgebinden; unpublished paper, Bielefeld.

Isermann, H. (1988):
Einsparung von Logistikkosten durch bessere Nutzung des Palettenstauraums; in: RKWHandbuch Logistik, 14.Lfg. XII/88, Berlin, pp. 1 - 28.

Isermann, H. (1991a):
Obere Schranken für die Lösung des zweidimensionalen Packproblems auf der Basis struktureller Identitäten; in: Fandel, G., Gehring, H. (eds.): Operations Research - Beiträge zur quantitativen Wirtschaftsforschung, Berlin et al., chapter 22, pp. 341 - 348.

Isermann, H. (1991b):
Verpackungsplanung im Spannungsfeld zwischen ökologischen und ökonomischen Anforderungen an die Verpackung; in: ORS 13, pp. 173 - 188.

Isermann, H. (1991c):
Heuristiken zur Lösung des zweidimensionalen Packproblems für Rundgefäße; in: ORS 13, pp. 213 - 223.

Israni, S., Sanders, J. (1982):
Two-Dimensional Cutting Stock Problem Research; in: Journal of Manufacturing Systems 1, pp. 169 - 182.

Israni, S. S., Sanders, J. L. (1984):
A Manufacturing Decision Support System for Flamecutting; in: Computers and Industrial Engineering 8, pp. 207 - 214.

Israni, S. S., Sanders, J. L. (1985):
Performance Testing of Rectangular Parts-Nesting Heuristics; in: IJPR 23, pp. 437 - 456.

Ivancic, N., Mathur, K., Mohanty, B. B. (1989):
An Integer Programming Based Heuristic Approach to the Three-Dimensional Packing Problem; in: Journal of Manufacturing and Operations Management 2, pp. 268 - 298.

Jackson, J. R. (1956):
A Computing Procedure for a Line Balancing Problem; in: MS 2, pp. 261 - 272.

Jansen, R., Graefenstein, T. (1988):
CAD-Systeme - Computer optimiert Verpackung; in: Jahrbuch der Logistik, pp. 98 - 101.

Johns, E. C. (1967):
Heuristic Procedures for Solving the Paper Trim Problem; in: Pierce, J. F. (ed.): Operations Research and the Design of Management Information Systems, TAPPI STAP Series 4, New York, chapter 20, pp. 361 - 369.

Johnson, D. S. (1974):
Fast Algorithms for Bin Packing; in: JCSS 8, pp. 272 - 314.

Johnson, D. S., Garey, M. R. (1985):
A 71/60 Theorem for Bin Packing; in: Journal of Complexity 1, pp. 65 - 106.

Johnson, E. L. (1989):
Modeling and Strong Linear Programs for Mixed Integer Programming; in: Wallace, S. W. (ed.): Algorithms and Model Formulations in Mathematical Programming, Berlin/Heidelberg, pp. 1 - 43.

Johnson, E. L., Padberg, M. W. (1981):
A Note on the Knapsack Problem with Special Ordered Sets; in: OR Letters 1, pp. 18 - 22.

Johnson, D. S., Demers, A., Ullman, J. D., Garey, M. R. , Graham, R. L. (1974):
Worst-Case Performance Bounds for Simple One-Dimensional Packing Algorithms; in: SIAM Computing 3, pp. 299 - 325.

Johnston, R. E. (1981):
OR in the Paper Industry; in: OMEGA 9, pp. 43 - 50.

Johnston, R. E. (1982):
A Direct Combinatorial Algorithm for Cutting Stock Problems; in: Anderssen, R. S., Hoog, F. R. de (eds.): The Application of Mathematics in Industry, Den Haag, pp. 137 - 152.

Johnston, R. E. (1986):
Rounding Algorithms for Cutting Stock Problems; in: Asia-Pacific Journal of Operational Research 3, pp. 166 - 171.

Johnston, R. E., Bourke, S. B. (1973):
The Development of Computer Programmes for Reel Deckle Filling; in: Appita 26, pp. 444 - 448.

Kämpke, T. (1988):
Simulated Annealing: Use of a New Tool in Bin Packing; in: Annals of OR 16, pp. 327 - 332.

Käschel, J., Mädler, A., Richter, K. (1991):
Software zur Verpackungslogistik - Übersicht und Vergleich; in: ORS 13, pp. 229 - 238.

Kästing, H. (1976):
Ein lexikographisches Suchverfahren zur Lösung des bivalenten Knapsack-Problems; in: Dathe, H. N. et al. (eds.): Proceedings in Operations Research 5, Würzburg, pp. 41 - 42.

Kallio, K. (1972):
The Trim Problem; in: Data (Scandinavian EDP-journal) 1, pp. 18 - 19.

Kallio, M. (1977):
Production Planning in a Paper Mill; in: Proceedings in Operations Research 7, Würzburg, pp. 278 - 284.

Kantorovich, L. V. (1960):
Mathematical Methods of Organizing and Planning Production; in: MS 6, pp. 366 - 422.

Karmarkar, N. (1982):
Probabilistic Analysis of Some Bin-Packing Problems; in: IEEE Transactions, pp. 107 - 111.

Karmarkar, N., Karp, R. M. (1982):
An Efficient Approximation Scheme for the One-Dimensional Bin-Packing Problem; in: IEEE Transactions, pp. 312 - 320.

Karnin, E. D. (1984):
A Parallel Algorithm for the Knapsack Problem; in: IEEE Transactions on Computers C-33, pp. 404 - 408.

Kaufman, L., Plastria, F., Tubeeckx, S. (1985):
The Zero-One Knapsack Problem with Equality Constraint; in: EJOR 19, pp. 384 - 389.

Keber, R. (1985):
Stauraumprobleme bei Stückguttransporten, Wissenschaftliche Berichte des Institutes für Fördertechnik der Universität Karlsruhe, Heft 17.

Klein, M. (1963):
On Assembly Line Balancing; in: OR 11, pp. 274 - 281.

Kleitman, D. J., Krieger, M. M. (1975):
An Optimal Bound for Two Dimensional Bin Packing; in: Proceedings of the 16th Annual Symposium on Foundations of Computing Science, IEEE, Long Beach, pp. 163 -168.

Klingst, A. (1965):
Rationelle Dimensionierung von Verpackungen; in: APF 3, pp. 337 - 358.

Knödel, W. (1981):
A Bin Packing Algorithnm with Complexity 0 (n log n) and Performance 1 in the Stochastic Limit, Mathematical Foundations of Computer Science; in: Gruska/Chytil (eds.): Lectional Notes in Computer Science 118, pp. 369 - 378.

Knödel, W. (1983):
Über das mittlere Verhalten von on-line Packungsalgorithmen; in: Elektronische Informationsverarbeitung und Kybernetik 19, pp. 427 - 433.

Köhler, H. (1978):
Das Überlängenproblem bei optimalen Zuschneideprogrammen einer Papierfabrik; in: WiSt 7, pp. 340 - 342.

Körner, D. (1974):
Lagenoptimierung als Zuschnittplanung im Bekleidungsbetrieb; in: BuW 19, pp. 1217 - 1221.

Kolesar, P. J. (1967):
A Branch and Bound Algorithm for the Knapsack Problem; in: MS 13, pp. 723 - 735.

Konno, H. (1981):
An Algorithm for Solving Bilinear Knapsack Problems; in: JORS of Japan 24, pp. 360 - 374.

Korchemkin, M. B. (1983):
A Heuristic Partitioning Algorithm for a Packaging Problem; in: Computing 31, pp. 203 - 210.

Kortanek, K., Sodaro, D. (1966):
A Generalized Network Model for Three-Dimensional Cutting Stock Problems and New Product Analysis; in: JIE 17, pp. 572 - 576.

Kou, L. T., Markowsky, G. (1977):
Multidimensional Bin Packing Algorithms; in: IBM Journal of Research and Development 21, pp. 443 - 448.

Krause, K., Larmore, L. L., Volper, D. J. (1987):
Packing Items from a Triangular Distribution; in: IPL 25, pp. 351 - 361.

Kreko, B. (1965):
Lehrbuch der Linearen Optimierung; 2nd edition, Berlin, in particular: "Das Zuschnittproblem"; pp. 286 - 287.

Kruse, H. J. (1991):
Entwicklung und Einsatz eines PC-Programms für das Problem der Palettenbeladung mit Pharmaprodukten via Kartonverpackung; in: ORS 13, pp. 242 - 248.

Kumara, S. R. T., Kashyap, R. L., Moodie, C. L. (1988):
Application of Expert Systems and Pattern Recognition Methodologies to Facilities Layout Planning; in: IJPR 26, pp. 905 - 930.

Lai, K. K., Chan, J. W. M. (1990a):
A Study of the Cutting Stock Problem, Technical Note AM-90-T01, Department of Applied Mathematics, City Polytechnic of Hong Kong, unpublished paper.

Lai, K. K., Chan, W. M. (1990b):
Two Dimensional Cutting Stock Algorithms: Bin Packing Approach, Research Report AM-90-07, Department of Applied Mathematics, City Polytechnic of Hong Kong, unpublished paper.

Lam, K. P. (1983):
A Hierarchical Method for Large-Scale Two-Dimensional Layout; in: Transactions of the ASME - Journal of Mechanisms, Transmissions, and Automation in Design 105, pp. 242 - 248.

Lambe, T. A. (1974):
Bounds on the Number of Feasible Solutions to a Knapsack Problem; in: SIAM Journal on Applied Mathematics 26, pp. 302 - 305.

Lampl, T., Stahl, J. (1965):
Über den optimalen Zuschnitt von Plattenmaterialien; in: Unternehmensforschung 9, pp. 187 - 197.

Langston, M. A. (1984):
Performance of Heuristics for a Computer Resource Allocation Problem; in: SIAM J.Alg.Dis.Meth. 5, pp. 154 - 161.

Langston, M. A. (1987):
A Study of Composite Heuristic Algorithms; in: JORS 38, pp. 539 - 544.

Larichev, O. I., Furems, E. M. (1986):
Multicriterion Packing Problem; in: Preprints of the 12th International Conference on MCDM, 1, pp. 388 - 399.

Larsen, O., Mikkelsen, G. (1980):
An Interactive System for the Loading of Cargo Aircraft; in: EJOR 4, pp. 367 - 373.

Lasdon, L. S. (1972):
Optimization Theory for Large Systems; 2nd edition, London, in particular: Chapter 4, "Solution of Linear Programs with Many Columns by Column- Generation Procedures"; pp. 207 - 266.

Laubenstein, T., Schneeweiß, C., Vaterrodt, H. J. (1982):
Verschnittoptimierung im praktischen Einsatz - Eine Fallstudie; in: ORS 4, pp. 229 - 236.

Laurent, D. G., Iyengar, S. S. (1982):
A Heuristic Algorithm for Optimal Placement of Rectangular Objects; in: Information Sciences 26, pp. 127 - 139.

Lauriere, M. (1978):
An Algorithm for the 0-1 Knapsack Problem; in: MP 14, pp. 1 - 10.

Lawler, E. L. (1979):
Fast Approximation Algorithms for Knapsack Problems; in: Mathematics of Operations Research 4, pp. 339 - 356.

Lee, D. H. (1979):
Optimal Loading of Tankers; in: JORS 30, pp. 323 - 329.

Lembersky, M. R., Chi, U. H. (1986):
Weyerhaeuser Decision Simulator Improves Timber Profits; in: Interfaces 16 (1), pp. 6 - 15.

Lev, B. (1972):
On the Loading Problem - A Comment; in: MS 18, pp. 428 - 431.

Levcopoulos, C., Lingas, A. (1984):
Covering Polygons with Minimum Number of Rectangles; in: Lecture Notes in Computer Science 166, pp. 63 - 72.

Lewis, R. T., Parker, R. G. (1982):
On a Generalized Bin-Packing Problem; in: NRLQ 29, pp. 119 - 145.

Liang, F. M. (1980):
A Lower Bound for On-Line Bin Packing; in: IPL 10, pp. 76 - 79.

Litton, C. D. (1977):
A Frequency Approach to the One-Dimensional Cutting Problem for Carpet Rolls; in: ORQ 28, pp. 927 - 938.

Liu, N. C., Chen, L. C. (1981):
A New Algorithm for Container Loading; in: Proceedings 5th International Computer Software and Applications Conference IEEE, pp. 292 - 299.

Lorie, J., Savage, L. J. (1955):
Three Problems in Rationing Capital; in: Journal of Business 28, pp. 229 - 239.

Loulou, R. (1984):
Probabilistic Behaviour of Optimal Bin-Packing Solutions; in: OR Letters 3, pp. 129 - 135.

Luderer, B., Singer, K. (1991):
Optimierung des Längsteilens von Blechen bei der Kernherstellung im Transformatorenbau; unpublished paper, Chemnitz.

Lueker, G. S. (1983):
Bin Packing with Items Uniformly Distributed over Intervals [a,b], in: Proceedings 24th Annual Symposium of Computer Science, Tucson, pp. 289 - 297.

Luss, H. (1983):
An Extended Model for the Optimal Sizing of Records; in: JORS 34, pp. 1099 - 1105.

Maculan, N. (1984):
Relaxation Lagrangienne: Le Problème Du Knapsack 0-1; in: Canadian Journal of Operational Research and Information Processing 21, pp. 315 - 327.

Madsen, O. B. G. (1979):
Glass Cutting in a Small Firm; in: MP 17, pp. 85 - 90.

Madsen, O. B. G. (1980):
References Concerning the Cutting Stock Problem; IMSOR Working Paper

Madsen, O. B. G. (1988):
An Application of Travelling-Salesman Routines to Solve Pattern-Allocation Problems in the Glass Industry; in: JORS 39, pp. 249 - 256.

Magazine, M. J., Oguz, O. (1981):
A Fully Polynomial Approximation Algorithm for the 0-1 Knapsack Problem; in: EJOR 8, pp. 270 - 273.

Magazine, M. J., Oguz, O. (1984):
A Heuristic Algorithm for the Multidimensional Zero-One Knapsack Problem; in: EJOR 16, pp. 319 - 326.

Magazine, M. J., Nemhauser, G. L., Trotter, L. E. (1975):
When the Greedy Solution Solves a Class of Knapsack Problems; in: OR 23, pp. 207 - 217.

Mangalousis, F., Biblis, E. J., White, C. R. (1981):
Optimum Solution for the Cut-To-Size Particleboard Panel Problem; in: Forest Products Journal 31 (12), pp. 37 - 44.

Mannchen, K. (1989):
Rechnergestützte Verfahren zur Bildung von Ladeeinheiten; Wissenschaftliche Berichte des Institutes für Fördertechnik der Universität Karlsruhe, Heft 28.

Marchetti-Spaccamela, A., Vercellis, C. (1987):
Efficient On-Line Algorithms for the Knapsack Problem; in: Ottman, T. (ed.): Automata, Languages, and Programming, Lecture Notes in Computer Science 267, Berlin, pp. 445 - 456.

Marconi, R. (1971):
Heuristic Method for Minimizing Trim Loss in the Paper Industry; in: IBM Technical Disclosure Bulletin 14, pp. 325 - 327.

Marcotte, O. (1985):
The Cutting Stock Problem and Integer Rounding; in: MP 33, pp. 82 - 92.

Marcotte, O. (1986):
An Instance of the Cutting Stock Problem for which the Rounding Property Does Not Hold; in: OR Letters 4, pp. 239 - 243.

Martel, C. U. (1985):
A Linear Time Bin-Packing Algorithm; in: OR Letters 4, pp. 189 - 192.

Martello, S., Toth, P. (1977a):
An Upper Bound for the Zero-One Knapsack Problem and a Branch and Bound Algorithm; in: EJOR 1, pp. 169 - 175.

Martello, S., Toth, P. (1977b):
Branch and Bound Algorithms for the Solution of the General Unidimensional Knapsack Problem; in: Roubens, M. (ed.): Advances in Operations Research, North-Holland, Amsterdam et al., pp. 295 - 301.

Martello, S., Toth, P. (1978a):
Algorithm 37 - Algorithm for the Solution of the 0-1 Single Knapsack Problem; in: Computing 21, pp. 81 - 86.

Martello, S., Toth, P. (1978b):
Algorithm 37 - Algorithm for the Solution of the 0-1 Single Knapsack Problem; in: Computing 21, pp. 81 - 86.

Martello, S., Toth, P. (1979):
The 0-1 Knapsack Problem; in: Christofides, N., Mingozzi, A., Toth, P., Sandi, C. (eds.): Combinatorical Optimization, Chichester et al., chapter 9, pp. 237 - 279.

Martello, S., Toth, P. (1980a):
Optimal and Canonical Solutions of the Change Making Problem; in: EJOR 4, pp. 322 -329.

Martello, S., Toth, P. (1980b):
Solution of the Zero-One Multiple Knapsack Problem; in: EJOR 4, pp. 276 - 283.

Martello, S., Toth, P. (1980c):
A Note on the Ingargiola-Korsh Algorithm for One-Dimensional Knapsack Problems; in: OR 28, pp. 1226 - 1227.

Martello, S., Toth, P. (1981a):
Heuristic Algorithms for the Multiple Knapsack Problem; in: Computing 27, pp. 93 - 112.

Martello, S., Toth, P. (1981b):
A Branch and Bound Algorithm for the Zero-One Multiple Knapsack Problem; in: Discrete Applied Mathematics 3, pp. 275 - 288.

Martello, S., Toth, P. (1984):
A Mixture of Dynamic Programming and Branch-and-Bound for the Subset-Sum Problem; in: MS 30, pp. 765 - 771.

Martello, S., Toth, P. (1985a):
Algorithm 632 - A Program for the 0-1 Multiple Knapsack Problem; in: ACM Transactions on Mathematical Systems 11, pp. 135 - 140.

Martello, S., Toth, P. (1985b):
Approximation Schemes for the Subset-Sum Problem: Survey and Experimental Analysis; in: EJOR 22, pp. 56 - 69.

Martello, S., Toth, P. (1987):
Algorithms for Knapsack Problems; in: Annals of Discrete Mathematics 31, pp. 213 - 258.

Martello, S., Toth, P. (1988):
A New Algorithm for the 0-1 Knapsack Problem; in: MS 34, pp. 633 - 644.

Martello, S., Toth, P. (1990):
Knapsack Problems - Algorithms and Computer Implementations, Chichester et al..

Martin, R. R., Stephenson, P. C. (1988):
Putting Objects into Boxes; in: CAD 20, pp. 506 - 514.

Maruyama, K., Chang, S. K., Tang, D. T. (1977):
A General Packing Algorithm for Multidimensional Resource Requirements; in: International Journal of Computer and Information Sciences 6, pp. 131 - 149.

MAS-GmbH (1987):
MASter-Trim - Eine Software-Linie zur computerunterstützten Produktionsfeinplanung und Schneideplanerstellung bei der Rollen- und Formatproduktion in der Papier- und Folienindustrie, Deisenhofen/München, pamphlet.

Meerendonk, H. W. van den, Schouten, J. H. (1962):
Die Schnittverluste bei der Herstellung von Wellpappe; in: Philips Technische Rundschau 4/5, pp. 133 - 142.

Meerendonk, H. W. van den, Kerbosch, J. A. G. M., Medema, P. , Schouten, J. H. (1963):
Some Computational Aspects of a Trimloss Problem; in: Statistica Neerlandica 17, pp. 37 - 47.

Meier, A., Naß, D. (1986):
Integrierte Fertigung beim Zuschnitt von Kunststoffteilen; in: ZwF 81, pp. 665 - 668.

Meier, G. (1978):
Heuristische Walztafelkombination für ein Blechwalzwerk; in: Späth, H. (ed.): Fallstudien Operations Research, Vol. 1, München/Wien, pp. 119 - 133.

Meier, H. (1975):
EDV-Lagenoptimierung in der Zuschneiderei; in: BuW 27, pp. 949 - 951.

Meir, A., Moser, L. (1968):
On Packing of Squares and Cubes; in: Journal of Combinatorial Theory 5, pp. 126 - 134.

Melzak, Z. A. (1966):
Infinite Packing of Disks; in: Canadian Journal of Mathematics 18, pp. 838 - 852.

Merle, G., Grütz, M. (1978):
Entwicklung eines heuristischen Verfahrens zur quantitativen Lösung des Verpackungsproblems im Versandhandel; in: Späth, H. (ed.): Fallstudien Operations Research, Vol. 2, München/Wien, pp. 108 - 130.

Metzger, R. W. (1958):
Elementary Mathematical Programming; New York; in particular: Chapter 8, "Stock Slitting"; pp. 200 - 210.

Moll, R. (1985):
Verfahren zur mehrzielorientierten Verschnittplanung in der Wellpappenindustrie; Bergisch Gladbach/Köln.

Morabito, R. N., Arenales, M. N. (1990):
An And-Or-Graph Approach for Two-Dimensional Cutting Problems; unpublished paper, Universidade de Sao Paulo, Brasil.

Morabito, R. N., Arenales, M. N. (1991):
On Solving Large Two-Dimensional Guillotine Cutting Problems; unpublished paper, Universidade de Sao Paulo, Brasil.

Müller-Merbach, H. (1973):
Operations Research, München; in particular: Chapter 4.8.5, "Ein Verschnittproblem"; pp. 171 - 173.

Müller-Merbach, H. (1978):
An Improved Upper Bound for the Zero-One Knapsack Problem - A Note on the Paper by Martello and Toth; in: EJOR 2, pp. 212 - 213.

Müller-Merbach, H. (1981):
Die Konstruktion von Input-Output-Modellen; in: Bergner, H. (ed.): Planung und Rechnungswesen in der Betriebswirtschaftslehre, pp. 19 - 113.

Murgolo, F. D. (1987):
An Efficient Approximation Scheme for Variable-Sized Bin Packing; in: SIAM Computing 16, pp. 149 - 161.

Murphy, R. A. (1986):
Some Computational Results on Real 0-1 Knapsack Problems; in: OR Letters 5, pp. 67 - 71.

Naujoks, G. (1990):
Neue Heuristiken und Strukturanalysen zum zweidimensionalen homogenen Packproblem; in: Kistner, K.P. et al. (eds.): OR Proceedings 1989, pp. 257 - 263.

Naujoks, G. (1991):
Ein neuer Ansatz zur Bestimmung theoretischer Obergrenzen für das zweidimensionale orthogonale homogene Packproblem, in: ORS 13, pp. 224 - 228.

Nauss, R. M. (1976):
An Efficient Algorithm for the 0-1 Knapsack Problem; in: MS 23, pp. 27 - 31.

Nauss, R. M. (1978):
The 0-1 Knapsack Problem with Multiple Choice Constraints; in: EJOR 2, pp. 125 - 131.

Nickels, W. (1988):
A Knowledge-Based System for Integrated Solving Cutting Stock Problems and Production Control in the Paper Industry; in: Mitra, G. (ed.): NATO ASI Series, F 48, Mathematical Models for Decision Support, Berlin et al., pp. 471 - 485.

Nickels, W. (1990):
Ein wissensbasiertes System zur Produktionsplanung und-steuerung in der Papierindustrie; Fortschritt-Berichte VDI, Reihe 20, No.24, Düsseldorf.

Niederhausen, H. P. (1977a):
Automatische Schnittplan-Optimierung; in: VDI-Berichte 277, Optimale Rohstoffnutzung - Eine Aufgabe für den Ingenieur, Düsseldorf, pp. 121 - 126.

Niederhausen, H. P. (1977b):
Programm zum automatischen Herstellen eines Schneidplans für beliebig gestaltete Werkstücke durch elektronische Datenverarbeitung; in: Schweißen und Schneiden 29, pp. 103 - 105.

Niederhausen, H. P., Reuter, S. (1978):
Schnittpläne optimieren - mit Erfahrung oder mit EDV?; in: Holz- und Kunststoffverarbeitung 13, pp. 852 - 863.

Oliveira, J. F., Ferreira, J. S. (1990):
An Improved Version of Wang's Algorithm for Two-Dimensional Cutting Problems; in: EJOR 44, pp. 256 - 266.

Olorunniwo, F. O. (1986):
Admissible Cutting Patterns in the Trim Problem; in: Engineering Optimization 10, pp. 125 - 138.

Ong, H. L., Magazine, M. J., Wee, T. S. (1984):
Probabilistic Analysis of Bin Packing Heuristics; in: OR 32, pp. 983 - 998.

Orlin, J. B. (1985):
Some Very Easy Knapsack/Partition Problems; in: OR 33, pp. 1154 - 1160.

Ozden, M. (1988):
A Solution Procedure for General Knapsack Problems with a Few Constraints; in: Comp&OR 15, pp. 145 - 155.

Padberg, M. W. (1979):
Covering, Packing and Knapsack Problems; in: Annals of Discrete Mathematics 4, pp. 265 - 287.

Padberg, M. W. (1980):
(1,k)-Configurations and Facets for Packing Problems; in: MP 18, pp. 94 - 99.

Page, E. (1975):
A Note on a Two-Dimensional Dynamic Programming Problem; in: ORQ 26, pp. 321 - 324.

Pandit, S. N. N. (1962):
The Loading Problem; in: OR 10, pp. 639 - 646.

Pasche, C. (1991):
Production planning; in: EJOR 50, pp. 27 - 36.

Paul, R. J. (1979):
A Production Scheduling Problem in the Glass-Container Industry; in: OR 27, pp. 290 - 302.

Paull, A. E. (1956):
Linear Programming: A Key to Optimum Newsprint Production; in: PPMCan 57, pp. 145 - 150.

Paull, A. E., Walter, J. R. (1955):
The Trim Problem: An Application of Linear Programming to the Manufacture of Newsprint Paper; in: Econometrica 23, p. 336.

Pegels, C. C. (1967a):
Heuristic Scheduling Models for Variants of the Two-Dimensional Cutting-Stock Problem; in: TAPPI 50, pp. 532 - 535.

Pegels, C. C. (1967b):
A Comparison of Scheduling Models for Corrugator Production; in: JIE 18, pp. 466 - 472.

Peleg, K. (1971):
Computerised Pallet Stacking; in: Packaging Technology, September, pp. 18 - 25.

Peleg, K., Orlowski, S. (1971):
Optimisation of Citrus Packaging; in: Packaging Technology, September, p. 17.

Peleg, K., Peleg, E. (1976):
Container Dimensions for Optimal Utilization of Storage and Transportation Space; in: CAD 8, pp. 175 - 180.

Pelzer, H., Ruth, K. A. (1966):
Steuerung des Stoffflusses und optimale Brammenaufteilung durch Einsatz einer elektronischen Rechenanlage; in: Stahl und Eisen 86, pp. 100 - 106.

Penington, R. A., Tanchoco, J. M. A. (1988):
Robotic Palletization of Multiple Box Sizes; in: IJPR 26, pp. 95 - 105.

Pentico, D. W. (1974):
The Assortment Problem with Probabilistic Demands; in: MS 21, pp. 286 - 290.

Pentico, D. W. (1976):
The Assortment Problem with Nonlinear Cost Functions; in: OR 24, pp. 1129 - 1142.

Pentico, D. W. (1988):
The Discrete Two-Dimensional Assortment Problem; in: OR 36, pp. 324 - 332.

Pfefferkorn, C. E. (1975):
A Heuristic Problem Solving Design System for Equipment or Furniture Layouts; in: CACM 18, pp. 286 - 297.

Philipson, R. H., Ravindran, A. (1979):
Application of Mathematical Programming to Metal Cutting; in: Mathematical Programming Study 11, pp. 116 - 134.

Picard, J. C., Queyranne, M. (1981):
On the One-Dimensional Space Allocation Problem; in: OR 29, pp. 371 - 391.

Pierce, J. F. (1964):
Some Large-Scale Production Scheduling Problems in the Paper Industry; Englewood Cliffs.

Pierce, J. F. (1967):
Trimming in Practice: Discussion; in: Pierce, J. F. (ed.): Operations Research and the Design of Management Information Systems, TAPPI STAP Series 4, New York, chapter 21, pp. 370 - 374.

Pierce, J. F. (1968):
Application of Combinatorial Programming to a Class of All-Zero-One Integer Programming Problems; in: MS 15, pp. 191 - 209.

Pierce, J. F. (1970):
Pattern Sequencing and Matching in Stock Cutting Operations; in: TAPPI 53, pp. 668 - 678.

Pinto, P. A., Dannenbring, D. G., Khumawala, B. M. (1981):
Branch and Bound and Heuristic Procedures for Assembly Line Balancing with Paralleling of Stations; in: IJPR 19, pp. 565 - 576.

Pirkul, H. (1987):
A Heuristic Procedure for the Multiconstraint Zero-One Knapsack Problem; in: NRLQ 34, pp. 161 - 172.

Plateau, G., Elkihel, M. (1985):
A Hybrid Method for the 0-1 Knapsack Problem; in: ORV 49, pp. 277 - 293.

Pnevmaticos, S. M., Mann, S. H. (1972):
Dynamic Programming in Tree Bucking; in: Forest Products Journal 22 (2), pp. 26 - 30.

Poirier, C. C. (1967):
Automated Data Processing for the Corrugated Box Plant; in: Pierce, J. F. (ed.): Operations Research and the Design of Management Information Systems, TAPPI STAP Series 4, New York, chapter 24, pp. 416 - 432.

Portmann, M. C. (1990):
An Efficient Algorithm for Container Loading; unpublished paper, Nancy.

Prodinger, H. (1985):
Einige Bemerkungen zu einer Arbeit von W. Knödel über das mittlere Verhalten von On-Line Packungsalgorithmen; in: Elektronische Informationsverarbeitung und Kybernetic 21, pp. 3 - 7.

Puls, F. M., Tanchoco, J. M. A. (1986):
Robotic Implementation of Pallet Loading Patterns; in: IJPR 24, pp. 635 - 645.

Qu, W., Sanders, J. L. (1987):
A Nesting Algorithm for Irregular Parts and Factors Affecting Trim Losses; in: IJPR 25, pp. 381 - 397.

Qu, W., Sanders, J. L. (1989):
Sequence Selection of Stock Sheets in Two-Dimensional Layout Problems; in: IJPR 27, pp. 1553 - 1571.

Queyranne, M. (1985):
Bounds for Assembly Line Balancing Heuristics; in: OR 33, pp. 1353 - 1359.

Rao, M. R. (1976):
On the Cutting Stock Problem; in: Journal of the Computer Society of India 7, pp. 35 - 39.

Rationalisierungs-Kuratorium der Deutschen Wirtschaft (RKW) (1981):
Modul-Empfehlung - Anpassung der Verpackungsabmessungen an Ladeeinheiten mit den Grundflächen 800 x 1200 mm, 1000 x 1200 mm, 600 x 800 mm; RGV-Schriften für die Verpackungspraxis, 5th edition, Eschborn.

Rayward-Smith, V. J., Shing, M. T. (1983):
Bin Packing; in: Bulletin of the IMA (UK) 19, pp. 142 - 146.

Reinders, M. P., Hendriks, T. H. B. (1989):
Lumberproduction Optimization; in: EJOR 42, pp. 243 - 253.

Reuter, S. (1980):
Automatisches Optimieren der Platinenanordnung in Werkzeugkonstruktionen mit Hilfe der EDV; in: Werkstatt und Betrieb 113, pp. 559 - 562.

Reymann, W. (1977):
Schnittoptimierung für flächen- und stangenförmiges Material; in: VDI-Berichte 277, Optimale Rohstoffnutzung - Eine Aufgabe für den Ingenieur, pp. 117 - 120.

Rhee, W. S. T. (1985):
Convergence of Optimal Stochastic Bin Packing; in: OR Letters 4, pp. 121 - 123.

Rhee, W. S. T. (1988):
Correction to "Probabilistic Analysis of the Next Fit Decreasing Algorithm for Bin-Packing"; in: OR Letters 7, p. 211.

Rhee, W. S. T., Talagrand, M. (1989):
Optimal Bin Packing with Items of Random Sizes II; in: SIAM Computing 18, pp. 139 - 151.

Riemer, W., Hartung, A. (1977):
On-Line-Optimierung für den Zuschnitt in der Textil- und Bekleidungsindustrie; in: IBM Nachrichten 27, pp. 289 - 295.

Rijckaert, M. J. (1980):
The Two-Dimensional Cutting Stock Problem; in: ORV 37, pp. 281 - 284.

Rijckaert, M. J. (1981):
Two- and Three-Dimensional Cutting Stock Problems with Different Stocks; in: ORV 43, pp. 145 - 253.

Rijckaert, M. J. (1986):
Heuristics and the Role of Microcomputers for the One-Dimensional Cutting Stock Problem; in: ORV 56, pp. 85 - 92.

Rinnooy Kan, A. H. G., De Wit, J. R., Wijmenga, R. T. (1987):
Nonorthogonal Two-Dimensional Cutting Patterns; in: MS 33, pp. 670 - 684.

Roberts, S. A. (1984):
Application of Heuristic Techniques to the Cutting-Stock for Worktops; in: JORS 35, pp. 369 - 377.

Robson, J. M. (1974):
Bounds for Some Functions Concerning Dynamic Storage Allocation; in: Journal of the ACM 21, pp. 491 - 499.

Robson, J. M. (1977):
Worst Case Fragmentation of First Fit and Best Fit Storage Allocation Strategies; in: CJ 20, pp. 242 - 244.

Rocha, T. G., Almeida, R. D. A., Moreno, A. D. O., Maculan, N. (1981a):
The Administration of Standard Length Telephone Cable Reels; in: Annals of Discrete Mathematics 11, pp. 109 - 123.

Rocha, T. G., Almeida, R. D. A., Moreno, A. D. O., Maculan, N. (1981b):
On the Solution of the Standard Length Telephone Cable Reels Problem; in: Brans, J. P. (ed.): Operational Research '81, Amsterdam, pp. 825 - 840.

Rockstroh, O. (1978):
Die Abstimmung der Verpackung mit Palette und Container; in: Rationalisierung 29, pp. 203 - 206.

Rode, M., Rosenberg, O. (1987):
An Analysis of Heuristic Trim-Loss Algorithms; in: Engineering Costs and Production Economics 12, pp. 71 - 78.

Romanovskii, I. V. (1969):
Solution of the Guillotine Cutting Problem by the Method of Processing a List of States; in: Cybernetics 5, pp. 123 - 125.

Roodman, G. M. (1986):
Near-Optimal Solutions to One-Dimensional Cutting Stock Problems; in: Comp&OR 13, pp. 713 - 719.

Rosenblatt, M. J., Sinuany-Stern, Z. (1989):
Generating the Discrete Efficient Frontier to the Capitel Budgeting Problem; in: OR 37, pp. 384 - 394.

Roth, K. F., Vaughan, R. C. (1978):
Ineffiency in Packing Squares with Unit Squares; in: Journal of Combinatorial Theory (A) 24, pp. 170 - 186.

Saaty, T. L., Alexander, J. M. (1975):
Optimization and the Geometry of Numbers: Packing and Covering; in: SIAM Review 17, pp. 475- 519.

Sadowski, W. (1959):
A few Remarks on the Assortment Problem; in: MS 6, pp. 13 - 24.

Sahni, S. (1975):
Approximate Algorithms for the 0/1 Knapsack Problem; in: JACM 22, pp. 115 - 124.

Salkin, H. M., Kluyer, C. A. de (1975):
The Knapsack Problem: A Survey; in: NRLQ 22, pp. 127 - 144.

Salveson, M. E. (1955):
The Assembly Line Balancing Problem; in: JIE, pp. 18 - 25.

Salzer, J. J. (1983):
Optimale Stapelmuster; in: Neue Verpackung, pp. 1248 - 1252.

Sarin, S. C. (1983a):
The Mixed Disc Packing Problem: Part I: Some Bounds on Density; in: IIE Transactions 16, pp. 37 - 45.

Sarin, S. C. (1983b):
Two-Dimensional Stock Cutting Problems and Solution Methodologies; in: Journal for Engineering for Industry 105, pp. 155 - 160.

Sarin, S. C., Ahn, S. (1983):
The Mixed Disc Packing Problem: Part II. An Interactive Optimization Procedure; in: IIE Transactions 15, pp. 91 - 98.

Sarker, B. R. (1988):
An Optimum Solution for One-Dimensional Slitting Problems: A Dynamic Approach; in: JORS 39, pp. 749 - 755.

Sarker, B. R., Shanthikumar, J. G. (1983):
A Generalized Approach for Serial or Parallel Line Balancing; in: IJPR 21, pp. 109 - 133.

Schachtel, F. (1958):
Wirtschaftliches Ausschneiden von Blechteilen, Berlin et al.

Scheid, F. (1960):
Some Packing Problems; in: American Mathematical Monthly 67, pp. 231 - 235.

Scheithauer, G. (1990):
Optimaler Zuschnitt trapezförmiger Teile; in: Wissenschaftliche Zeitschrift der Technischen Universität Dresden 39, S. 191 - 195.

Scheithauer, G., Terno, J. (1985):
Lösung von Guillotine-Zuschnittproblemen mittels revidierter Simplexmethode und Spaltengenerierung; in: Beitrag zum 30. Internationalen Kolloquium, TH Ilmenau, pp. 135 - 138.

Scheithauer, G., Terno, J. (1986a):
Zerlegung eines Quadrates in flächengleiche Rechtecke; Informationen der TU Dresden; unpublished paper.

Scheithauer, G., Terno, J. (1986b):
Effektive Lösung von Guillotine-Zuschnittproblemen; in: Wissenschaftliche Zeitschrift der TU Dresden 35 (4), pp. 35 - 40.

Scheithauer, G., Terno, J. (1987):
An Effective Branch and Bound Approach to the Cutting Stock Problem; Informationen der TU Dresden; unpublished paper.

Scheithauer, G., Terno, J. (1988a):
The Solution of Guillotine Cutting Problems; unpublished paper, Dresden.

Scheithauer, G., Terno, J. (1988b):
The Partition of a Square in Rectangles with Equal Areas; in: Journal of Information Processing and Cybernetics 24, pp. 189 - 200.

Scheithauer, G., Terno, J. (1988c):
Guillotine Cutting of Defective Boards; in: Optimization 19, pp. 111 - 121.

Scheithauer, G., Terno, J. (1990):
A Heuristic Procedure for the Container Loading Problem; in: Informationen der TU Dresden, unpublished paper.

Schepens, G. (1978):
The Development and Implementation of an Interactive Cutting Model for the Corrugated Board Industry; in: Hax, A. C. (ed.): Studies in Operations Management, Amsterdam et al., chapter 10, pp. 286 - 301.

Schlag, M., Liao, Y. Z., Wong, C. K. (1983):
An Algorithm for Optimal Two-Dimensional Compaction of VLSI Layouts; in: Integration 1, pp. 179 - 209.

Schlingensiepen, J. (1987):
Kalkulation von Kosten für flexible Bearbeitungszentren; in: krp 4, pp. 135 - 144.

Schneider, W. (1979):
Ein Verfahren zur Lösung des 3-dimensionalen Verschnittproblems, unter Berücksichtigung schneidetechnischer und produktionstechnischer Restriktionen; Linz, doctoral thesis.

Schneider, W. (1987):
Zweidimensionale Verschnittminimierung - Ein Vergleich zwischen optimaler Lösung und Heuristik; in: Isermann, H. et al. (eds.): OR Proceedings 1986, Berlin et al., pp. 322 - 328.

Schneider, W. (1988):
Trim-Loss Minimization in a Crepe-Rubber Mill; Optimal Solution versus Heuristic in the 2 (3) - Dimensional Case; in: EJOR 34, pp. 273 - 281.

Schuster, K. P. (1991):
Logistikgerechte Konzeption sowie Realisierung der Produktverpackung und ein praktischer Anwendungsfall; in: ORS 13, pp. 254 - 263.

SCS-GmbH (1986):
Walztafelkombination in einem Grobblechwalzwerk - Zweidimensionale Verschnittprobleme, SCS Projekt Beschreibung, pamphlet.

Sculli, D. (1981):
A Stochastic Cutting Stock Procedure: Cutting Rolls of Insulating Tape; in: MS 27, pp. 946 - 952.

Sculli, D., Hui, C. F. (1988):
Three Dimensional Stacking of Containers; in: OMEGA 16, pp. 585 - 594.

Seth, A. (1987):
Wastage Reduction in Wood Cutting; in: Opsearch 24, pp. 94 - 105.

Seth, A., Prasad, V. R., Ramamurthy, K. G. (1986):
A Heuristic Approach to One-Dimensional Cutting Stock Problem; in: Opsearch 23, pp. 235 - 243.

Shapiro, J. F. (1968):
Dynamic Programming Algorithms for the Integer Programming Problem - I: The Integer Programming Problem Viewed as a Knapsack Type Problem; in: OR 16, pp. 103 - 121.

Shapiro, J. F., Wagner, H. M. (1967):
A Finite Renewal Algorithm for the Knapsack and Turnpike Models; in: OR 15, pp. 319 - 341.

Shapiro, S. D. (1977):
Performance of Heuristic Bin Packing Algorithms with Segments of Random Length; in: Information and Control 35, pp. 146 - 158.

Shearer, J. B. (1981):
A Counterexample to a Bin Packing Conjecture; in: SIAM J.Alg.Dis.Meth. 2, pp. 309 - 310.

Shearn, D. C. S. (1976):
The Assortment Problem: Different Models and an Application; in: ORQ 27, pp. 567 - 571.

Shih, W. (1979):
A Branch and Bound Method for the Multiconstraint Zero-One Knapsack Problem; in: JORS 30, pp. 369 - 378.

Shor, P. W. (1984):
The Average-Case Analysis of Some On-Line Algorithms for Bin Packing; in: Proceedings of the 25th Annual Symposium on Foundation of Computer Science, Singer Ivland, pp. 193 - 200.

Shor, P. W. (1986):
The Average-Case Analysis of Some On-Line Algorithms for Bin Packing; in: Combinatorica 6, pp. 179 - 200.

Simmons, D. M. (1972):
Optimal Inventory Policies under a Hierarchy of Setup Costs; in: MS 18, pp. 591 - 599.

Sinha, P., Zoltners, A. A. (1979):
The Multiple-Choice Knapsack Problem; in: OR 27, pp. 503 - 515.

Skalbeck, B. A., Schultz, H. K. (1976):
Reducing Trim Waste in Panel Cutting Using Integer and Linear Programming; in: Proceedings of the Western AIDS Conference, pp. 145 - 147.

Sleator, D. D. K. D. B. (1980):
A 2.5 Times Optimal Algorithm for Packing in Two Dimensions; in: IPL 10, pp. 37 - 40.

Smilauer, A. (1962):
Vereinfachte Berechnungen des Minimalabfalles; in: Messen - Steuern - Regeln 5, pp. 485 - 488.

Smith, A., Cani, P. de (1980):
An Algorithm to Optimize the Layout of Boxes in Pallets; in: JORS 31, pp. 573 - 578.

Smithin, T., Harrison, P. (1982):
The Third Dimension of Two-Dimensional Cutting; in: OMEGA 10, pp. 81 - 87.

Stadtler, H. (1983):
Stowing Seagoing Chemical Tankers: An Example of Solving some Semi-Structured Decision Problems; in: EJOR 14, pp. 279 - 287.

Stadtler, H. (1988):
A Comparison of Two Optimization Procedures for 1- and 1,5- Dimensional Cutting Stock Problems; in: ORS 10, pp. 97 - 111.

Stadtler, H. (1990):
A One-Dimensional Cutting Stock Problem in the Aluminium Industry and its Solution; in: EJOR 44, pp. 209 - 223.

Stainton, R. S. (1977):
The Cutting Stock Problem for the Stockholder of Steel Reinforcement Bars; in: ORQ 28, pp. 139 - 149.

Stehling, F. (1983):
Der Bedarf an Münzen in verschiedenen Münzsystemen; in: ORV 46, pp. 509 - 521.

Stern, H. I., Avivi, Z. (1990):
The Selection and Scheduling of Textile Orders with Due Dates; in: EJOR 44, pp. 11 - 16.

Steuckart, H. (1974):
Untersuchung von Vergleichskriterien für die Optimierung von Zuschneideverfahren, Forschungsberichte des Landes NRW, Nr. 2308, Köln-Opladen.

Steudel, H. J. (1979):
Generating Pallet Loading Patterns: A Special Case of the Two-Dimensional Cutting Stock Problem; in: MS 25, pp. 997 - 1004.

Steudel, H. J. (1984):
Generating Pallet Loading Patterns with Considerations of Item Stacking on End and Side Surfaces; in: Journal of Manufacturing Systems 3, pp. 135 - 143.

Stockmeyer, L. (1983):
Optimal Orientations of Cells in Slicing Floorplan Designs; in: Information and Control 57, pp. 91 - 101.

Stommel, H. J., Buschhoff, U. (1986):
Rechnergesteuerter Fertigungsablauf bei der Herstellung von Blechteilen; in: ZwF 81, pp. 624 - 628.

Stoyan, Y. G. (1983):
Mathematical Methods for Geometry Design; in: Ellis, T. M. R., Semenkoc, O. J. (eds.): Advances in CAD/CAM, Amsterdam, pp. 67 - 86.

Struve, D. L. (1967):
Paper Machine Production Allocation by Linear Programming; in: Pierce, J. F. (ed.): Operations Research and the Design of Management Information Systems, TAPPI STAP Series 4, New York, chapter 16, pp. 293 - 307.

Suhl, U. (1978):
An Algorithm and Efficient Data Structures for the Binary Knapsack Problem; in: EJOR 2, pp. 420 - 428.

Sumichrast, R. T. (1986):
A New Cutting-Stock Heuristic for Scheduling Production; in: Comp&OR 13, pp. 403 - 410.

Sweeney, P. E., Haessler, R. W. (1990):
One-Dimensional Cutting Stock Decisions for Rolls with Multiple Quality Grades; in: EJOR 44, pp. 224 - 231.

Sweeney, P. E., Ridenour, E. L. (1989):
Cutting and Packing Problems: A Categorized, Application-Oriented Research Bibliography; Working Paper #610, unpublished paper, Michigan.

Tabucanon, M. T., Lertcharoensombat, C. (1986):
Modeling the Trimming Problem in a Paper Production System; in: Geering, H. P., Mansour, M. (eds.): Large Scale Systems: Theory and Application 2, Pergamon, Oxford, pp. 509 - 513.

Talbot, F. B., Patterson, J. H., Gehrlein, W. V. (1986):
A Comparative Evaluation of Heuristic Line Balancing Techniques; in: MS 32, pp. 430 - 454.

Tanchoco, J. M. A., Agee, M. H. (1981):
Plan Units Loads to Interact with All Components of Warehouse System; in: Industrial Engineering, June, pp. 36 - 48.

Tanchoco, J. M. A., Agee, M. H., Davis, R. P., Wysk, R. A. (1980):
An Analysis of the Interactions between Unit Loads, Handling Equipment, Storage, and Shipping; in: Unit and Bulk Materials Handling, an ASME Publication Consisting of Papers Presented at the ASME Materials Handling Conference, August 19-21, San Francisco, pp. 185 - 190.

Tanchoco, J. M. A., Davis, R. P., Egbelu, P. J., Wysk, R. A. (1983):
Economic Unit Loads (EUL) for Multi-Product Inventory Systems with Limited Storage Space; in: Material Flow 1, pp. 141 - 148.

Terno, J., Lindemann, R., Scheithauer, G. (1987):
Zuschnittprobleme und ihre praktische Lösung; Frankfurt am Main.

Thesen, A. (1975):
A Recursive Branch and Bound Algorithm for the Multidimensional Knapsack Problem; in: NRLQ 22, pp. 341 - 353.

Tien, B. N., Hu, T. C. (1977):
Error Bounds and the Applicability of the Greedy Solution to the Coin-Changing Problem; in: OR 25, pp. 404 - 418.

Tilanus, C. B., Gerhardt, C. (1972):
Das Knapsack-Problem, Anmerkungen und Erweiterungen; in: OR 2, pp. 105 - 120.

Tilanus, C. B., Gerhardt, C. (1976):
An Application of Cutting Stock in the Steel Industry; in: Haley, K. B. (ed.): Operational Research '75, Amsterdam, pp. 669 - 675.

Toczylowski. E. (1986):
On Aggregation in a Two-Dimensional Cutting Stock Scheduling Problem; in: Large Scale Systems 10, pp. 165 - 174.

Tokuyama, H., Ueno, N. (1981):
The Cutting Stock Problems in the Iron and Steel Industries; in: Brans, J. P. (ed.): Operational Research '81, Amsterdam, pp. 809 - 823.

Tokuyama, H., Ueno, N. (1985):
The Cutting Stock Problem for Large Sections in the Iron and Steel Industries; in: EJOR 22, pp. 280 - 292.

Toth, P. (1980):
Dynamic Programming Algorithms for the Zero-One Knapsack Problem; in: Computing 25, pp. 29 - 45.

Troßmann, E. (1983):
Verschnittoptimierung dargestellt an Beispielen aus der Textilindustrie; Berlin.

Tsai, R. D., Malstrom, E. M., Meeks, H. D. (1988):
A Two-Dimensional Palletizing Procedure for Warehouse Loading Operations; in: IIE Transactions 20, pp. 418 - 425.

Vajda, S. (1958):
Trim Loss Reduction; in: Readings in Mathematical Programming, 2nd edition, London, chapter 21, pp. 78 - 84.

Vasko, F. J. (1989):
A Computational Improvement to Wang's Two-Dimensional Cutting Stock Algorithm; in: Computers and Industrial Engineering 16, pp. 109 - 115.

Vasko, F. J., Wolf, F. E., Stott, K. L. (1989):
A Practical Solution to a Fuzzy Two-Dimensional Cutting Stock Problem; in: Fuzzy Sets and Systems 29, pp. 259 - 275.

Vasko, F. J., Wolf, F. E., Pflugrad, J. A. (1991):
An Efficient Heuristic for Planning Mother Plate Requirements at Bethlehem Steel; in: Interfaces 21 (2), pp. 1 - 7.

VEB Wissenschaftlich-Technisches Zentrum der holzverarbeitenden Industrie (1988): SOPHIA - Schnittplan-Optimierung plattenförmiger Werkstoffe der holzverarbeitenden Industrie und anderer Industriezweige; Dresden, pamphlet.

Vel, H. van de, Shijie, S. (1991): An Application of the Bin-packing Technique to Job Scheduling on Uniform Processors; in: JORS 42, pp. 169 - 172.

Veliev, G. P., Mamedov, K. S. (1981): A Method of Solving the Knapsack Problem; in: USSR Comput. Maths. Phys. 21 (3), pp. 75 - 81.

Voigt, J. U. (1987): Solving Cutting Stock Problems of the Furniture Industry via Man-Machine-Communication; in: Wissentschaftliche Zeitschrift der TH Ilmenau 33 (6), pp. 113 - 118.

Vonderembse, M. A. (1979): An Investigation of Design and Operating Decisions for Continuous Casting in the Steel Industry; Michigan, doctoral thesis.

Vonderembse, M. A., Haessler, R. W. (1982): A Mathematical Programming Approach to Schedule Master Slab Casters in the Steel Industry; in: MS 28, pp. 1450 - 1461.

Wäscher, G. (1989a): Neuere Entwicklungen auf dem Gebiet der Zuschnittplanung; in: Pressmar, D. et al. (eds.): OR Proceedings 1988, Berlin et al., pp. 469 - 475.

Wäscher, G. (1989b): Zuschnittplanung - Probleme, Modelle, Lösungsverfahren; postdoctoral thesis, Stuttgart.

Wäscher, G. (1990): An LP-Based Approach to Cutting Stock Problems with Multiple Objectives; in: EJOR 44, pp. 175 - 184.

Wäscher, G., Müller, H. (1986): Developing a Computer Program for Cutting Problems in a Steel Rolling Mill; in: Systems Analytical Modelling Simulation 3, pp. 321 - 330.

Wäscher, G., Carow, P., Müller, H. (1985): Entwicklung eines flexiblen Verfahrens für Zuschneideprobleme in einem Kaltwalzwerk; in: ZOR 29, pp. B 209 - B 230.

Walters, J. R. (1976a):
Approximate Stock Formulae for Use in Conjunction with the Assortment Problem; in: ORQ 27, pp. 801 - 804.

Walters, J. R. (1976b):
An Extension to the Assortment Problem in which Components are Grouped to Form Assemblies; in: ORQ 27, pp. 169 - 175.

Walther, E. C. (1991):
Optimierung von Transportverpackungen in der PBS-Industrie; in: ORS 13, pp. 239 - 241.

Walukiewicz, S. (1975):
The Size Reduction of a Binary Knapsack Problem; in: Bulletin de L'Academie Polonaise des Sciences, Serie des Sciences Techniques 23, pp. 453 - 458.

Wang, P. Y. (1983):
Two Algorithms for Constrained Two-Dimensional Cutting Stock Problems; in: OR 31, pp. 573 - 586.

Wedekind, H. (1968):
Ein Verfahren zur Bestimmung der maximalen Lagenzahl beim Zuschneiden in der Konfektionsindustrie; in: Bussmann, K. F., Mertens, P. (eds.): OR und DV bei der Produktionsplanung; Stuttgart, pp. 197 - 215.

Wee, T. S., Magazine, M. J. (1982):
Assembly Line Balancing as Generalized Bin Packing; in: OR Letters 1, pp. 56 - 58.

Weingartner, H. M., Ness, D. N. (1967):
Methods for the Solution of the Multi-Dimensional 0/1 Knapsack Problem; in: OR 15, pp. 83 - 103.

Weiser, C. (1988):
Computer Utilization in Packaging; in: Journal of Packaging Technology 2, pp. 203 - 204.

Whitaker, D., Cammell, S. (1990):
A Partitioned Cutting-Stock Problem Applied in the Meat Industry; in: JORS 41, pp. 801 - 807.

Wihl, W. (1969):
Verschnittminimierung beim Zuschnitt aus Platten in Standardgrößen; in: Bürotechnik und Automation 10, pp. 186 - 192.

Wilson, R. C. (1965):
A Packaging Problem; in: MS 12, pp. 135 - 145.

Wingerter, L. (1975):
Computer Programs Cut Years off Package-Design Leadtime; in: Modern Materials Handling, pp. 46 - 48.

Wintgen, G., Kluge, H. (1961):
Über den optimalen Zuschnitt von Blechen, ein Problem der Linearprogrammierung; in: Wissenschaftliche Zeitschrift der Technischen Universität Dresden 10, pp. 1291 - 1294.

Wirsam, B. (1987):
Köpfchen gegen Computer - Zuschnittoptimierung von Spanplatten in Theorie und Praxis; in: Holz-Zentralblatt, p. 28.

Wolfson, M. L. (1965):
Selecting the Best Lengths to Stock; in: OR 13, pp. 570 - 585.

Woolsey, R. E. D. (1972):
A Candle to Saint Jude, or Four Real World Applications of Integer Programming; in: Interfaces 2 (2), pp. 20 - 27.

Wright, G. (1979):
There May Be a Better Pack for You; in: The Australian Lithographer 12 (71), pp. 33 - 34.

Wright, P. G. (1973):
A Systems Approach to Packaging Design; in: Australian Packaging, June, pp. 27 - 33.

Wright, P. G. (1974):
Pallet Loading Configurations for Optimal Storage and Shipping; in: Paperboard Packaging, pp. 46 - 49.

Yanasse, H. H., Soma, N. Y. (1987):
A New Enumeration Scheme for the Knapsack Problem; in: Discrete Applied Mathematics 18, pp. 235 - 245.

Yanasse, H. H., Zinober, A. S. I., Harris, R. G. (1990):
Cutting Stock: Board Size Selection; unpublished paper, University of Sheffield.

Yanasse, H. H., Zinober, A. S. I., Harris, R. G. (1991):
Two-dimensional Cutting Stock with Multiple Stock Sizes; in: JORS 42, pp. 673 - 683.

Yao, A. C. C. (1980):
New Algorithms for Bin Packing; in: JACM 27, pp. 207 - 227.

Yeong, W. Y., Yue, T. M. (1991):
Cutting Down Trim-loss for Competitive Advantage: a Singaporean Company's Experience; in: JORS 42, pp. 649 - 654.

Zaloom, V. A. (1982):
An Automated Procedure to Establish Workzone Boundaries for Air Force Facilities Maintenance Operations; in: JORS 33, pp. 913 - 919.

Zemel, E. (1980):
The Linear Multiple Choice Knapsack Problem; in: OR 28, pp. 1412 - 1423.

Zemel, E. (1984):
An O(n) Algorithm for the Linear Multiple Choice Knapsack Problem and Related Problems; in: IPL 18, pp. 123 - 128.

Zimmermann, H. J. (1971):
Einführung in die Grundlagen des Operations Research, München, in particular: chapter 2.1.5., "Verschnittmodelle"; pp. 160 - 163.

Zimmermann, E. (1975):
Gradierung und Schnittbild-Optimierung mit EDV-Anlagen; in: BuW 27, pp. 726 - 728.

Zissinopoulos, V. (1985):
Heuristic Methods for Solving (Un)Constrained Two-Dimensional Cutting Stock Problems; in: ORV 49, pp. 345 - 357.

Zoltners, A. A. (1978):
A Direct Descent Binary Knapsack Algorithm; in: JACM 25, pp. 304 - 311

MIX
Papier aus verantwortungsvollen Quellen
Paper from responsible sources
FSC® C105338

If you have any concerns about our products,
you can contact us on
ProductSafety@springernature.com

In case Publisher is established outside the EU,
the EU authorized representative is:
Springer Nature Customer Service Center GmbH
Europaplatz 3, 69115 Heidelberg, Germany

Printed by Libri Plureos GmbH
in Hamburg, Germany